"After utilizing toolkits from The Art of Service, I was able to identify threats within my organization to which I was completely unaware. Using my team's knowledge as a competitive advantage, we now have superior systems that save time and energy."

"As a new Chief Technology Officer, I was feeling unprepared and inadequate to be successful in my role. I ordered an IT toolkit Sunday night and was prepared Monday morning to shed light on areas of improvement within my organization. I no longer felt overwhelmed and intimidated, I was excited to share what I had learned."

"I used the questionnaires to interview members of my team. I never knew how many insights we could produce collectively with our internal knowledge."

"I usually work until at least 8pm on weeknights. The Art of Service questionnaire saved me so much time and worry that Thursday night I attended my son's soccer game without sacrificing my professional obligations."

"After purchasing The Art of Service toolkit, I was able to identify areas where my company was not in compliance that could have put my job at risk. I looked like a hero when I proactively educated my team on the risks and presented a solid solution."

"I spent months shopping for an external consultant before realizing that The Art of Service would allow my team to consult themselves! Not only did we save time not catching a consultant up to speed, we were able to keep our company information and industry secrets confidential."

"Everyday there are new regulations and processes in my industry. The Art of Service toolkit has kept me ahead by using AI technology to constantly update the toolkits and address emerging needs."

"I customized The Art of Service toolkit to focus specifically on the concerns of my role and industry. I didn't have to waste time with a generic self-help book that wasn't tailored to my exact situation."

"Many of our competitors have asked us about our secret sauce. When I tell them it's the knowledge we have in-house, they never believe me. Little do they know The Art of Service toolkits are working behind the scenes."

"One of my friends hired a consultant who used the knowledge gained working with his company to advise their competitor. Talk about a competitive disadvantage! The Art of Service allowed us to keep our knowledge from walking out the door along with a huge portion of our budget in consulting fees."

"Honestly, I didn't know what I didn't know. Before purchasing The Art of Service, I didn't realize how many areas of my business needed to be refreshed and improved. I am so relieved The Art of Service was there to highlight our blind spots."

"Before The Art of Service, I waited eagerly for consulting company reports to come out each month. These reports kept us up to speed but provided little value because they put our competitors on the same playing field. With The Art of Service, we have uncovered unique insights to drive our business forward."

"Instead of investing extensive resources into an external consultant, we can spend more of our budget towards pursuing our company goals and objectives...while also spending a little more on corporate holiday parties."

"The risk of our competitors getting ahead has been mitigated because The Art of Service has provided us with a 360-degree view of threats within our organization before they even arise."

Autonomous car liability
Complete Self-Assessment Guide

https://theartofservice.com
support@theartofservice.com

Table of Contents

About The Art of Service 9

Included Resources - how to access 9
Purpose of this Self-Assessment 11
How to use the Self-Assessment 12
Autonomous car liability
Scorecard Example 14
Autonomous car liability
Scorecard 15

BEGINNING OF THE
SELF-ASSESSMENT: 16
CRITERION #1: RECOGNIZE 17

CRITERION #2: DEFINE: 28

CRITERION #3: MEASURE: 44

CRITERION #4: ANALYZE: 58

CRITERION #5: IMPROVE: 74

CRITERION #6: CONTROL: 91

CRITERION #7: SUSTAIN: 104
Autonomous car liability and Managing Projects, Criteria for
Project Managers: 129
1.0 Initiating Process Group: Autonomous car liability 130

1.1 Project Charter: Autonomous car liability 132

1.2 Stakeholder Register: Autonomous car liability 134

1.3 Stakeholder Analysis Matrix: Autonomous car liability 135

2.0 Planning Process Group: Autonomous car liability 137

2.1 Project Management Plan: Autonomous car liability 139

2.2 Scope Management Plan: Autonomous car liability 141

2.3 Requirements Management Plan: Autonomous car liability 144

2.4 Requirements Documentation: Autonomous car liability 146

2.5 Requirements Traceability Matrix: Autonomous car liability 148

2.6 Project Scope Statement: Autonomous car liability 150

2.7 Assumption and Constraint Log: Autonomous car liability 152

2.8 Work Breakdown Structure: Autonomous car liability 154

2.9 WBS Dictionary: Autonomous car liability 156

2.10 Schedule Management Plan: Autonomous car liability 159

2.11 Activity List: Autonomous car liability 161

2.12 Activity Attributes: Autonomous car liability 163

2.13 Milestone List: Autonomous car liability 165

2.14 Network Diagram: Autonomous car liability 167

2.15 Activity Resource Requirements: Autonomous car liability 169

2.16 Resource Breakdown Structure: Autonomous car liability 171

2.17 Activity Duration Estimates: Autonomous car liability 173

2.18 Duration Estimating Worksheet: Autonomous car liability 175

2.19 Project Schedule: Autonomous car liability 177

2.20 Cost Management Plan: Autonomous car liability 179

2.21 Activity Cost Estimates: Autonomous car liability 181

2.22 Cost Estimating Worksheet: Autonomous car liability 183

2.23 Cost Baseline: Autonomous car liability 185

2.24 Quality Management Plan: Autonomous car liability 187

2.25 Quality Metrics: Autonomous car liability 189

2.26 Process Improvement Plan: Autonomous car liability 191

2.27 Responsibility Assignment Matrix: Autonomous car liability 193

2.28 Roles and Responsibilities: Autonomous car liability 195

2.29 Human Resource Management Plan: Autonomous car liability 197

2.30 Communications Management Plan: Autonomous car liability 199

2.31 Risk Management Plan: Autonomous car liability 201

2.32 Risk Register: Autonomous car liability 203

2.33 Probability and Impact Assessment: Autonomous car liability 205

2.34 Probability and Impact Matrix: Autonomous car liability 207

2.35 Risk Data Sheet: Autonomous car liability 209

2.36 Procurement Management Plan: Autonomous car liability 211

2.37 Source Selection Criteria: Autonomous car liability 213

2.38 Stakeholder Management Plan: Autonomous car liability 215

2.39 Change Management Plan: Autonomous car liability 217

3.0 Executing Process Group: Autonomous car liability 219

3.1 Team Member Status Report: Autonomous car liability 221

3.2 Change Request: Autonomous car liability 223

3.3 Change Log: Autonomous car liability 225

3.4 Decision Log: Autonomous car liability 227

3.5 Quality Audit: Autonomous car liability 229

3.6 Team Directory: Autonomous car liability 231

3.7 Team Operating Agreement: Autonomous car liability 233

3.8 Team Performance Assessment: Autonomous car liability 235

3.9 Team Member Performance Assessment: Autonomous car liability 237

3.10 Issue Log: Autonomous car liability 239

4.0 Monitoring and Controlling Process Group: Autonomous car liability 241

4.1 Project Performance Report: Autonomous car liability 243

4.2 Variance Analysis: Autonomous car liability 245

4.3 Earned Value Status: Autonomous car liability 247

4.4 Risk Audit: Autonomous car liability 249

4.5 Contractor Status Report: Autonomous car liability 251

4.6 Formal Acceptance: Autonomous car liability 253

5.0 Closing Process Group: Autonomous car liability 255

5.1 Procurement Audit: Autonomous car liability 257

5.2 Contract Close-Out: Autonomous car liability 259

5.3 Project or Phase Close-Out: Autonomous car liability 261

5.4 Lessons Learned: Autonomous car liability 263
Index 265

About The Art of Service

The Art of Service, Business Process Architects since 2000, is dedicated to helping stakeholders achieve excellence.

Defining, designing, creating, and implementing a process to solve a stakeholders challenge or meet an objective is the most valuable role... In EVERY group, company, organization and department.

Unless you're talking a one-time, single-use project, there should be a process. Whether that process is managed and implemented by humans, AI, or a combination of the two, it needs to be designed by someone with a complex enough perspective to ask the right questions.

Someone capable of asking the right questions and step back and say, 'What are we really trying to accomplish here? And is there a different way to look at it?'

With The Art of Service's Self-Assessments, we empower people who can do just that — whether their title is marketer, entrepreneur, manager, salesperson, consultant, Business Process Manager, executive assistant, IT Manager, CIO etc... —they are the people who rule the future. They are people who watch the process as it happens, and ask the right questions to make the process work better.

Contact us when you need any support with this Self-Assessment and any help with templates, blue-prints and examples of standard documents you might need:

https://theartofservice.com
support@theartofservice.com

Included Resources - how to access

Included with your purchase of the book is the Autonomous

car liability Self-Assessment Spreadsheet Dashboard which contains all questions and Self-Assessment areas and auto-generates insights, graphs, and project RACI planning - all with examples to get you started right away.

How? Simply send an email to
access@theartofservice.com
with this books' title in the subject to get the Autonomous car liability Self Assessment Tool right away.

The auto reply will guide you further, you will then receive the following contents with New and Updated specific criteria:

- The latest quick edition of the book in PDF
- The latest complete edition of the book in PDF, which criteria correspond to the criteria in...
- The Self-Assessment Excel Dashboard, and...
- Example pre-filled Self-Assessment Excel Dashboard to get familiar with results generation
- In-depth specific Checklists covering the topic
- Project management checklists and templates to assist with implementation

INCLUDES LIFETIME SELF ASSESSMENT UPDATES

Every self assessment comes with Lifetime Updates and Lifetime Free Updated Books. Lifetime Updates is an industry-first feature which allows you to receive verified self assessment updates, ensuring you always have the most accurate information at your fingertips.

Get it now- you will be glad you did - do it now, before you forget.

Send an email to **access@theartofservice.com** with this books' title in the subject to get the Autonomous car liability Self Assessment Tool right away.

Purpose of this Self-Assessment

This Self-Assessment has been developed to improve understanding of the requirements and elements of Autonomous car liability, based on best practices and standards in business process architecture, design and quality management.

It is designed to allow for a rapid Self-Assessment to determine how closely existing management practices and procedures correspond to the elements of the Self-Assessment.

The criteria of requirements and elements of Autonomous car liability have been rephrased in the format of a Self-Assessment questionnaire, with a seven-criterion scoring system, as explained in this document.

In this format, even with limited background knowledge of Autonomous car liability, a manager can quickly review existing operations to determine how they measure up to the standards. This in turn can serve as the starting point of a 'gap analysis' to identify management tools or system elements that might usefully be implemented in the organization to help improve overall performance.

How to use the Self-Assessment

On the following pages are a series of questions to identify to what extent your Autonomous car liability initiative is complete in comparison to the requirements set in standards.

To facilitate answering the questions, there is a space in front of each question to enter a score on a scale of '1' to '5'.

1 Strongly Disagree

2 Disagree

3 Neutral

4 Agree

5 Strongly Agree

Read the question and rate it with the following in front of mind:

'In my belief,
the answer to this question is clearly defined'.

There are two ways in which you can choose to interpret this statement;

1. how aware are you that the answer to the question is clearly defined
2. for more in-depth analysis you can choose to gather evidence and confirm the answer to the question. This obviously will take more time, most Self-Assessment users opt for the first way to interpret the question and dig deeper later on based on the outcome of the overall Self-Assessment.

A score of '1' would mean that the answer is not clear at all, where a '5' would mean the answer is crystal clear and defined. Leave emtpy when the question is not applicable or you don't want to answer it, you can skip it without affecting your score. Write your score in the space provided.

After you have responded to all the appropriate statements in each section, compute your average score for that section, using the formula provided, and round to the nearest tenth. Then transfer to the corresponding spoke in the Autonomous car liability Scorecard on the second next page of the Self-Assessment.

Your completed Autonomous car liability Scorecard will give you a clear presentation of which Autonomous car liability areas need attention.

Autonomous car liability Scorecard Example

Example of how the finalized Scorecard can look like:

Autonomous car liability Scorecard

Your Scores:

BEGINNING OF THE SELF-ASSESSMENT:

CRITERION #1: RECOGNIZE

INTENT: Be aware of the need for change. Recognize that there is an unfavorable variation, problem or symptom.

In my belief, the answer to this question is clearly defined:

5 Strongly Agree

4 Agree

3 Neutral

2 Disagree

1 Strongly Disagree

1. What extra resources will you need?
<--- Score

2. How do you take a forward-looking perspective in identifying Autonomous car liability research related to market response and models?
<--- Score

3. What else needs to be measured?

<--- Score

4. What creative shifts do you need to take?
<--- Score

5. Will a response program recognize when a crisis occurs and provide some level of response?
<--- Score

6. Which issues are too important to ignore?
<--- Score

7. Is the need for organizational change recognized?
<--- Score

8. What situation(s) led to this Autonomous car liability Self Assessment?
<--- Score

9. Which needs are not included or involved?
<--- Score

10. Are there Autonomous car liability problems defined?
<--- Score

11. Who needs what information?
<--- Score

12. To what extent does each concerned units management team recognize Autonomous car liability as an effective investment?
<--- Score

13. How do you recognize an objection?
<--- Score

14. What Autonomous car liability events should you attend?
<--- Score

15. What should be considered when identifying available resources, constraints, and deadlines?
<--- Score

16. How are training requirements identified?
<--- Score

17. What prevents you from making the changes you know will make you a more effective Autonomous car liability leader?
<--- Score

18. Do you recognize Autonomous car liability achievements?
<--- Score

19. Are there regulatory / compliance issues?
<--- Score

20. What is the problem and/or vulnerability?
<--- Score

21. What needs to be done?
<--- Score

22. Can management personnel recognize the monetary benefit of Autonomous car liability?
<--- Score

23. Do you know what you need to know about Autonomous car liability?

<--- Score

24. How do you recognize an Autonomous car liability objection?
<--- Score

25. What are the clients issues and concerns?
<--- Score

26. How do you identify the kinds of information that you will need?
<--- Score

27. Are your goals realistic? Do you need to redefine your problem? Perhaps the problem has changed or maybe you have reached your goal and need to set a new one?
<--- Score

28. What is the recognized need?
<--- Score

29. What is the extent or complexity of the Autonomous car liability problem?
<--- Score

30. Are problem definition and motivation clearly presented?
<--- Score

31. What resources or support might you need?
<--- Score

32. What do you need to start doing?
<--- Score

33. For your Autonomous car liability project, identify and describe the business environment, is there more than one layer to the business environment?
<--- Score

34. What would happen if Autonomous car liability weren't done?
<--- Score

35. Are you dealing with any of the same issues today as yesterday? What can you do about this?
<--- Score

36. What tools and technologies are needed for a custom Autonomous car liability project?
<--- Score

37. How much are sponsors, customers, partners, stakeholders involved in Autonomous car liability? In other words, what are the risks, if Autonomous car liability does not deliver successfully?
<--- Score

38. How many trainings, in total, are needed?
<--- Score

39. Are employees recognized or rewarded for performance that demonstrates the highest levels of integrity?
<--- Score

40. Would you recognize a threat from the inside?
<--- Score

41. To what extent would your organization benefit from being recognized as a award recipient?

<--- Score

42. Are losses recognized in a timely manner?
<--- Score

43. Where is training needed?
<--- Score

44. Who needs to know?
<--- Score

45. What is the smallest subset of the problem you can usefully solve?
<--- Score

46. Will new equipment/products be required to facilitate Autonomous car liability delivery, for example is new software needed?
<--- Score

47. What problems are you facing and how do you consider Autonomous car liability will circumvent those obstacles?
<--- Score

48. Are employees recognized for desired behaviors?
<--- Score

49. How do you assess your Autonomous car liability workforce capability and capacity needs, including skills, competencies, and staffing levels?
<--- Score

50. Are there recognized Autonomous car liability problems?
<--- Score

51. How are you going to measure success?
<--- Score

52. Did you miss any major Autonomous car liability issues?
<--- Score

53. What Autonomous car liability problem should be solved?
<--- Score

54. What Autonomous car liability capabilities do you need?
<--- Score

55. Are there any specific expectations or concerns about the Autonomous car liability team, Autonomous car liability itself?
<--- Score

56. What is the Autonomous car liability problem definition? What do you need to resolve?
<--- Score

57. What are the timeframes required to resolve each of the issues/problems?
<--- Score

58. How are the Autonomous car liability's objectives aligned to the group's overall stakeholder strategy?
<--- Score

59. Who should resolve the Autonomous car liability issues?
<--- Score

60. Does Autonomous car liability create potential expectations in other areas that need to be recognized and considered?
<--- Score

61. Does your organization need more Autonomous car liability education?
<--- Score

62. As a sponsor, customer or management, how important is it to meet goals, objectives?
<--- Score

63. Which information does the Autonomous car liability business case need to include?
<--- Score

64. What is the problem or issue?
<--- Score

65. How do you identify subcontractor relationships?
<--- Score

66. What activities does the governance board need to consider?
<--- Score

67. Is it needed?
<--- Score

68. Why is this needed?
<--- Score

69. Have you identified your Autonomous car liability key performance indicators?

<--- Score

70. What are the stakeholder objectives to be achieved with Autonomous car liability?
<--- Score

71. Think about the people you identified for your Autonomous car liability project and the project responsibilities you would assign to them, what kind of training do you think they would need to perform these responsibilities effectively?
<--- Score

72. What do employees need in the short term?
<--- Score

73. Who else hopes to benefit from it?
<--- Score

74. Where do you need to exercise leadership?
<--- Score

75. Whom do you really need or want to serve?
<--- Score

76. What needs to stay?
<--- Score

77. Who needs budgets?
<--- Score

78. Will Autonomous car liability deliverables need to be tested and, if so, by whom?
<--- Score

79. Is the quality assurance team identified?

<--- Score

80. What are the Autonomous car liability resources needed?
<--- Score

81. What vendors make products that address the Autonomous car liability needs?
<--- Score

82. What are the expected benefits of Autonomous car liability to the stakeholder?
<--- Score

83. Who defines the rules in relation to any given issue?
<--- Score

84. What does Autonomous car liability success mean to the stakeholders?
<--- Score

85. Who are your key stakeholders who need to sign off?
<--- Score

86. What are your needs in relation to Autonomous car liability skills, labor, equipment, and markets?
<--- Score

87. Is it clear when you think of the day ahead of you what activities and tasks you need to complete?
<--- Score

88. How can auditing be a preventative security measure?

<--- Score

89. What Autonomous car liability coordination do you need?
<--- Score

90. Do you need different information or graphics?
<--- Score

91. What training and capacity building actions are needed to implement proposed reforms?
<--- Score

92. Do you need to avoid or amend any Autonomous car liability activities?
<--- Score

93. Who needs to know about Autonomous car liability?
<--- Score

Add up total points for this section:
_____ = Total points for this section

Divided by: ______ (number of statements answered) = ______ Average score for this section

Transfer your score to the Autonomous car liability Index at the beginning of the Self-Assessment.

CRITERION #2: DEFINE:

INTENT: Formulate the stakeholder problem. Define the problem, needs and objectives.

In my belief, the answer to this question is clearly defined:

5 Strongly Agree

4 Agree

3 Neutral

2 Disagree

1 Strongly Disagree

1. What information should you gather?
<--- Score

2. Has/have the customer(s) been identified?
<--- Score

3. Are resources adequate for the scope?
<--- Score

4. What sort of initial information to gather?
<--- Score

5. Is the team adequately staffed with the desired cross-functionality? If not, what additional resources are available to the team?
<--- Score

6. What is a worst-case scenario for losses?
<--- Score

7. What sources do you use to gather information for a Autonomous car liability study?
<--- Score

8. What would be the goal or target for a Autonomous car liability's improvement team?
<--- Score

9. Is there any additional Autonomous car liability definition of success?
<--- Score

10. Are there different segments of customers?
<--- Score

11. Is the improvement team aware of the different versions of a process: what they think it is vs. what it actually is vs. what it should be vs. what it could be?
<--- Score

12. What are the Autonomous car liability use cases?
<--- Score

13. Does the scope remain the same?
<--- Score

14. Is Autonomous car liability linked to key stakeholder goals and objectives?
<--- Score

15. Are customer(s) identified and segmented according to their different needs and requirements?
<--- Score

16. What are (control) requirements for Autonomous car liability Information?
<--- Score

17. Do you all define Autonomous car liability in the same way?
<--- Score

18. Is scope creep really all bad news?
<--- Score

19. Who are the Autonomous car liability improvement team members, including Management Leads and Coaches?
<--- Score

20. Are accountability and ownership for Autonomous car liability clearly defined?
<--- Score

21. How do you gather the stories?
<--- Score

22. When is/was the Autonomous car liability start date?
<--- Score

23. How will variation in the actual durations of each activity be dealt with to ensure that the expected Autonomous car liability results are met?
<--- Score

24. What defines best in class?
<--- Score

25. How have you defined all Autonomous car liability requirements first?
<--- Score

26. Has anyone else (internal or external to the group) attempted to solve this problem or a similar one before? If so, what knowledge can be leveraged from these previous efforts?
<--- Score

27. How do you keep key subject matter experts in the loop?
<--- Score

28. Are approval levels defined for contracts and supplements to contracts?
<--- Score

29. Is it clearly defined in and to your organization what you do?
<--- Score

30. Are required metrics defined, what are they?
<--- Score

31. Have all of the relationships been defined properly?
<--- Score

32. How would you define the culture at your organization, how susceptible is it to Autonomous car liability changes?
<--- Score

33. Is the scope of Autonomous car liability defined?
<--- Score

34. Are there any constraints known that bear on the ability to perform Autonomous car liability work? How is the team addressing them?
<--- Score

35. Have all basic functions of Autonomous car liability been defined?
<--- Score

36. How will the Autonomous car liability team and the group measure complete success of Autonomous car liability?
<--- Score

37. What Autonomous car liability services do you require?
<--- Score

38. What intelligence can you gather?
<--- Score

39. Are roles and responsibilities formally defined?
<--- Score

40. What are the boundaries of the scope? What is in bounds and what is not? What is the start point? What is the stop point?

<--- Score

41. How do you manage unclear Autonomous car liability requirements?
<--- Score

42. Is there a clear Autonomous car liability case definition?
<--- Score

43. What Autonomous car liability requirements should be gathered?
<--- Score

44. What baselines are required to be defined and managed?
<--- Score

45. Is there a completed, verified, and validated high-level 'as is' (not 'should be' or 'could be') stakeholder process map?
<--- Score

46. Is the work to date meeting requirements?
<--- Score

47. How can the value of Autonomous car liability be defined?
<--- Score

48. Is there a completed SIPOC representation, describing the Suppliers, Inputs, Process, Outputs, and Customers?
<--- Score

49. How did the Autonomous car liability manager

receive input to the development of a Autonomous car liability improvement plan and the estimated completion dates/times of each activity?
<--- Score

50. How and when will the baselines be defined?
<--- Score

51. What is the context?
<--- Score

52. What happens if Autonomous car liability's scope changes?
<--- Score

53. Has the direction changed at all during the course of Autonomous car liability? If so, when did it change and why?
<--- Score

54. How do you manage changes in Autonomous car liability requirements?
<--- Score

55. In what way can you redefine the criteria of choice clients have in your category in your favor?
<--- Score

56. What is the scope of Autonomous car liability?
<--- Score

57. How are consistent Autonomous car liability definitions important?
<--- Score

58. What customer feedback methods were used to

solicit their input?
<--- Score

59. Has a high-level 'as is' process map been completed, verified and validated?
<--- Score

60. What is out-of-scope initially?
<--- Score

61. Why are you doing Autonomous car liability and what is the scope?
<--- Score

62. What critical content must be communicated – who, what, when, where, and how?
<--- Score

63. Where can you gather more information?
<--- Score

64. How do you hand over Autonomous car liability context?
<--- Score

65. Do the problem and goal statements meet the SMART criteria (specific, measurable, attainable, relevant, and time-bound)?
<--- Score

66. What are the compelling stakeholder reasons for embarking on Autonomous car liability?
<--- Score

67. What are the core elements of the Autonomous car liability business case?

<--- Score

68. What gets examined?
<--- Score

69. What specifically is the problem? Where does it occur? When does it occur? What is its extent?
<--- Score

70. What system do you use for gathering Autonomous car liability information?
<--- Score

71. Is data collected and displayed to better understand customer(s) critical needs and requirements.
<--- Score

72. What are the requirements for audit information?
<--- Score

73. How do you build the right business case?
<--- Score

74. The political context: who holds power?
<--- Score

75. Has everyone on the team, including the team leaders, been properly trained?
<--- Score

76. What is the definition of Autonomous car liability excellence?
<--- Score

77. If substitutes have been appointed, have they

been briefed on the Autonomous car liability goals and received regular communications as to the progress to date?
<--- Score

78. What is the scope of the Autonomous car liability effort?
<--- Score

79. Are task requirements clearly defined?
<--- Score

80. What is in scope?
<--- Score

81. Who defines (or who defined) the rules and roles?
<--- Score

82. Who is gathering Autonomous car liability information?
<--- Score

83. Are different versions of process maps needed to account for the different types of inputs?
<--- Score

84. What is the scope of the Autonomous car liability work?
<--- Score

85. What are the Autonomous car liability tasks and definitions?
<--- Score

86. Is the Autonomous car liability scope complete and appropriately sized?

<--- Score

87. How would you define Autonomous car liability leadership?
<--- Score

88. Is special Autonomous car liability user knowledge required?
<--- Score

89. When are meeting minutes sent out? Who is on the distribution list?
<--- Score

90. Who approved the Autonomous car liability scope?
<--- Score

91. Is Autonomous car liability required?
<--- Score

92. Has a Autonomous car liability requirement not been met?
<--- Score

93. Are the Autonomous car liability requirements complete?
<--- Score

94. What information do you gather?
<--- Score

95. What is in the scope and what is not in scope?
<--- Score

96. Is the Autonomous car liability scope manageable?

<--- Score

97. When is the estimated completion date?
<--- Score

98. Do you have a Autonomous car liability success story or case study ready to tell and share?
<--- Score

99. Scope of sensitive information?
<--- Score

100. Has a team charter been developed and communicated?
<--- Score

101. How do you catch Autonomous car liability definition inconsistencies?
<--- Score

102. Does the team have regular meetings?
<--- Score

103. What are the dynamics of the communication plan?
<--- Score

104. What is the definition of success?
<--- Score

105. What are the rough order estimates on cost savings/opportunities that Autonomous car liability brings?
<--- Score

106. Do you have organizational privacy

requirements?
<--- Score

107. How do you think the partners involved in Autonomous car liability would have defined success?
<--- Score

108. Has the improvement team collected the 'voice of the customer' (obtained feedback – qualitative and quantitative)?
<--- Score

109. Have specific policy objectives been defined?
<--- Score

110. What are the Roles and Responsibilities for each team member and its leadership? Where is this documented?
<--- Score

111. Is the current 'as is' process being followed? If not, what are the discrepancies?
<--- Score

112. What constraints exist that might impact the team?
<--- Score

113. Has your scope been defined?
<--- Score

114. Has a project plan, Gantt chart, or similar been developed/completed?
<--- Score

115. What scope to assess?

<--- Score

116. How is the team tracking and documenting its work?
<--- Score

117. Is Autonomous car liability currently on schedule according to the plan?
<--- Score

118. Are audit criteria, scope, frequency and methods defined?
<--- Score

119. What are the record-keeping requirements of Autonomous car liability activities?
<--- Score

120. Is there a Autonomous car liability management charter, including stakeholder case, problem and goal statements, scope, milestones, roles and responsibilities, communication plan?
<--- Score

121. How do you gather requirements?
<--- Score

122. What knowledge or experience is required?
<--- Score

123. How often are the team meetings?
<--- Score

124. Is there regularly 100% attendance at the team meetings? If not, have appointed substitutes attended to preserve cross-functionality and full

representation?
<--- Score

125. How was the 'as is' process map developed, reviewed, verified and validated?
<--- Score

126. What scope do you want your strategy to cover?
<--- Score

127. Have the customer needs been translated into specific, measurable requirements? How?
<--- Score

128. How do you manage scope?
<--- Score

129. How does the Autonomous car liability manager ensure against scope creep?
<--- Score

130. Are all requirements met?
<--- Score

131. Is there a critical path to deliver Autonomous car liability results?
<--- Score

132. What are the tasks and definitions?
<--- Score

133. Who is gathering information?
<--- Score

134. Has the Autonomous car liability work been fairly and/or equitably divided and delegated among team

members who are qualified and capable to perform the work? Has everyone contributed?
<--- Score

135. What key stakeholder process output measure(s) does Autonomous car liability leverage and how?
<--- Score

136. What is the worst case scenario?
<--- Score

Add up total points for this section:
_____ = Total points for this section

Divided by: ______ (number of statements answered) = ______ Average score for this section

Transfer your score to the Autonomous car liability Index at the beginning of the Self-Assessment.

CRITERION #3: MEASURE:

INTENT: Gather the correct data. Measure the current performance and evolution of the situation.

In my belief, the answer to this question is clearly defined:

5 Strongly Agree

4 Agree

3 Neutral

2 Disagree

1 Strongly Disagree

1. Are actual costs in line with budgeted costs?
<--- Score

2. How frequently do you track Autonomous car liability measures?
<--- Score

3. Are the Autonomous car liability benefits worth its costs?

<--- Score

4. Is there an opportunity to verify requirements?
<--- Score

5. How to cause the change?
<--- Score

6. Where is the cost?
<--- Score

7. How will measures be used to manage and adapt?
<--- Score

8. Which costs should be taken into account?
<--- Score

9. How will costs be allocated?
<--- Score

10. Are you able to realize any cost savings?
<--- Score

11. Have design-to-cost goals been established?
<--- Score

12. What drives O&M cost?
<--- Score

13. How will your organization measure success?
<--- Score

14. Do you effectively measure and reward individual and team performance?
<--- Score

15. What are allowable costs?
<--- Score

16. What causes investor action?
<--- Score

17. What could cause you to change course?
<--- Score

18. What are the costs of delaying Autonomous car liability action?
<--- Score

19. How do you measure success?
<--- Score

20. What users will be impacted?
<--- Score

21. What potential environmental factors impact the Autonomous car liability effort?
<--- Score

22. When a disaster occurs, who gets priority?
<--- Score

23. How will effects be measured?
<--- Score

24. What is the cost of rework?
<--- Score

25. Is the cost worth the Autonomous car liability effort ?
<--- Score

26. What tests verify requirements?
<--- Score

27. Who should receive measurement reports?
<--- Score

28. What are the strategic priorities for this year?
<--- Score

29. What are your primary costs, revenues, assets?
<--- Score

30. Why do you expend time and effort to implement measurement, for whom?
<--- Score

31. How is the value delivered by Autonomous car liability being measured?
<--- Score

32. What would it cost to replace your technology?
<--- Score

33. Why do the measurements/indicators matter?
<--- Score

34. How is performance measured?
<--- Score

35. Who pays the cost?
<--- Score

36. Does a Autonomous car liability quantification method exist?
<--- Score

37. What are the current costs of the Autonomous car liability process?
<--- Score

38. How do you verify performance?
<--- Score

39. What are the Autonomous car liability key cost drivers?
<--- Score

40. Are there any easy-to-implement alternatives to Autonomous car liability? Sometimes other solutions are available that do not require the cost implications of a full-blown project?
<--- Score

41. What are the estimated costs of proposed changes?
<--- Score

42. What would be a real cause for concern?
<--- Score

43. Do the benefits outweigh the costs?
<--- Score

44. What is your decision requirements diagram?
<--- Score

45. What is an unallowable cost?
<--- Score

46. What evidence is there and what is measured?
<--- Score

47. How can you measure the performance?
<--- Score

48. Do you have an issue in getting priority?
<--- Score

49. Where can you go to verify the info?
<--- Score

50. How do you measure variability?
<--- Score

51. What does your operating model cost?
<--- Score

52. What is the Autonomous car liability business impact?
<--- Score

53. Is it possible to estimate the impact of unanticipated complexity such as wrong or failed assumptions, feedback, etcetera on proposed reforms?
<--- Score

54. How are measurements made?
<--- Score

55. Are you aware of what could cause a problem?
<--- Score

56. Is the solution cost-effective?
<--- Score

57. What are the manufacturers obligations with respect to crashes caused by Cyberattacks?

<--- Score

58. Are supply costs steady or fluctuating?
<--- Score

59. Does the Autonomous car liability task fit the client's priorities?
<--- Score

60. What are the costs?
<--- Score

61. How do you verify your resources?
<--- Score

62. What methods are feasible and acceptable to estimate the impact of reforms?
<--- Score

63. When should you bother with diagrams?
<--- Score

64. What are your key Autonomous car liability organizational performance measures, including key short and longer-term financial measures?
<--- Score

65. How do you measure efficient delivery of Autonomous car liability services?
<--- Score

66. Are the measurements objective?
<--- Score

67. What are the uncertainties surrounding estimates of impact?

<--- Score

68. Have you included everything in your Autonomous car liability cost models?
<--- Score

69. What can be used to verify compliance?
<--- Score

70. What are hidden Autonomous car liability quality costs?
<--- Score

71. How much does it cost?
<--- Score

72. Are Autonomous car liability vulnerabilities categorized and prioritized?
<--- Score

73. How can you manage cost down?
<--- Score

74. What are the costs and benefits?
<--- Score

75. How can you measure Autonomous car liability in a systematic way?
<--- Score

76. How do you prevent mis-estimating cost?
<--- Score

77. Does management have the right priorities among projects?
<--- Score

78. How do you verify if Autonomous car liability is built right?
<--- Score

79. How do you measure lifecycle phases?
<--- Score

80. What are the operational costs after Autonomous car liability deployment?
<--- Score

81. Are missed Autonomous car liability opportunities costing your organization money?
<--- Score

82. What is your Autonomous car liability quality cost segregation study?
<--- Score

83. How do you quantify and qualify impacts?
<--- Score

84. How do you control the overall costs of your work processes?
<--- Score

85. Will Autonomous car liability have an impact on current business continuity, disaster recovery processes and/or infrastructure?
<--- Score

86. Among the Autonomous car liability product and service cost to be estimated, which is considered hardest to estimate?
<--- Score

87. What disadvantage does this cause for the user?
<--- Score

88. What are your operating costs?
<--- Score

89. How is progress measured?
<--- Score

90. Are you taking your company in the direction of better and revenue or cheaper and cost?
<--- Score

91. What are your customers expectations and measures?
<--- Score

92. How long to keep data and how to manage retention costs?
<--- Score

93. What does a Test Case verify?
<--- Score

94. How do you verify the authenticity of the data and information used?
<--- Score

95. How sensitive must the Autonomous car liability strategy be to cost?
<--- Score

96. What is the total cost related to deploying Autonomous car liability, including any consulting or professional services?

<--- Score

97. How can you reduce the costs of obtaining inputs?
<--- Score

98. What measurements are being captured?
<--- Score

99. How do your measurements capture actionable Autonomous car liability information for use in exceeding your customers expectations and securing your customers engagement?
<--- Score

100. How do you verify the Autonomous car liability requirements quality?
<--- Score

101. How do you aggregate measures across priorities?
<--- Score

102. What details are required of the Autonomous car liability cost structure?
<--- Score

103. Was a business case (cost/benefit) developed?
<--- Score

104. Do you have a flow diagram of what happens?
<--- Score

105. What happens if cost savings do not materialize?
<--- Score

106. What harm might be caused?

<--- Score

107. What does losing customers cost your organization?
<--- Score

108. Are indirect costs charged to the Autonomous car liability program?
<--- Score

109. How can you reduce costs?
<--- Score

110. How will success or failure be measured?
<--- Score

111. What are the costs of reform?
<--- Score

112. How can a Autonomous car liability test verify your ideas or assumptions?
<--- Score

113. At what cost?
<--- Score

114. Which Autonomous car liability impacts are significant?
<--- Score

115. Which measures and indicators matter?
<--- Score

116. What relevant entities could be measured?
<--- Score

117. What causes mismanagement?
<--- Score

118. Are there competing Autonomous car liability priorities?
<--- Score

119. What could cause delays in the schedule?
<--- Score

120. Where is it measured?
<--- Score

121. Have you made assumptions about the shape of the future, particularly its impact on your customers and competitors?
<--- Score

122. What are the types and number of measures to use?
<--- Score

123. What is the root cause(s) of the problem?
<--- Score

124. What are the Autonomous car liability investment costs?
<--- Score

125. How do you verify and validate the Autonomous car liability data?
<--- Score

126. What are you verifying?
<--- Score

127. What measurements are possible, practicable and meaningful?
<--- Score

128. Did you tackle the cause or the symptom?
<--- Score

129. What is the total fixed cost?
<--- Score

130. How will you measure success?
<--- Score

131. Do you have any cost Autonomous car liability limitation requirements?
<--- Score

132. Do you aggressively reward and promote the people who have the biggest impact on creating excellent Autonomous car liability services/products?
<--- Score

133. What causes extra work or rework?
<--- Score

Add up total points for this section:
_____ = Total points for this section

Divided by: ______ (number of statements answered) = ______ Average score for this section

Transfer your score to the Autonomous car liability Index at the beginning of the Self-Assessment.

CRITERION #4: ANALYZE:

INTENT: Analyze causes, assumptions and hypotheses.

In my belief, the answer to this question is clearly defined:

5 Strongly Agree

4 Agree

3 Neutral

2 Disagree

1 Strongly Disagree

1. What do you need to qualify?
<--- Score

2. Think about the functions involved in your Autonomous car liability project, what processes flow from these functions?
<--- Score

3. What other organizational variables, such as reward systems or communication systems, affect the

performance of this Autonomous car liability process?
<--- Score

4. Is the required Autonomous car liability data gathered?
<--- Score

5. What Autonomous car liability data should be collected?
<--- Score

6. Is the performance gap determined?
<--- Score

7. How will the Autonomous car liability data be captured?
<--- Score

8. What qualifications do Autonomous car liability leaders need?
<--- Score

9. Who qualifies to gain access to data?
<--- Score

10. How is data used for program management and improvement?
<--- Score

11. What are the revised rough estimates of the financial savings/opportunity for Autonomous car liability improvements?
<--- Score

12. Record-keeping requirements flow from the records needed as inputs, outputs, controls and for

transformation of a Autonomous car liability process, are the records needed as inputs to the Autonomous car liability process available?
<--- Score

13. What will drive Autonomous car liability change?
<--- Score

14. Think about some of the processes you undertake within your organization, which do you own?
<--- Score

15. What kind of crime could a potential new hire have committed that would not only not disqualify him/her from being hired by your organization, but would actually indicate that he/she might be a particularly good fit?
<--- Score

16. What were the crucial 'moments of truth' on the process map?
<--- Score

17. Do staff qualifications match your project?
<--- Score

18. How do mission and objectives affect the Autonomous car liability processes of your organization?
<--- Score

19. What successful thing are you doing today that may be blinding you to new growth opportunities?
<--- Score

20. Who gets your output?

<--- Score

21. How do you implement and manage your work processes to ensure that they meet design requirements?
<--- Score

22. How do you ensure that the Autonomous car liability opportunity is realistic?
<--- Score

23. Was a cause-and-effect diagram used to explore the different types of causes (or sources of variation)?
<--- Score

24. Is the Autonomous car liability process severely broken such that a re-design is necessary?
<--- Score

25. What quality tools were used to get through the analyze phase?
<--- Score

26. What Autonomous car liability data will be collected?
<--- Score

27. Do your leaders quickly bounce back from setbacks?
<--- Score

28. What qualifies as competition?
<--- Score

29. How does the organization define, manage, and improve its Autonomous car liability processes?

<--- Score

30. What are the necessary qualifications?
<--- Score

31. How much data can be collected in the given timeframe?
<--- Score

32. Have you defined which data is gathered how?
<--- Score

33. What does the data say about the performance of the stakeholder process?
<--- Score

34. Did any additional data need to be collected?
<--- Score

35. What tools were used to narrow the list of possible causes?
<--- Score

36. What qualifications are necessary?
<--- Score

37. What methods do you use to gather Autonomous car liability data?
<--- Score

38. Is there any way to speed up the process?
<--- Score

39. How do you measure the operational performance of your key work systems and processes, including productivity, cycle time, and other appropriate

measures of process effectiveness, efficiency, and innovation?
<--- Score

40. What are your current levels and trends in key Autonomous car liability measures or indicators of product and process performance that are important to and directly serve your customers?
<--- Score

41. Do you, as a leader, bounce back quickly from setbacks?
<--- Score

42. What are your best practices for minimizing Autonomous car liability project risk, while demonstrating incremental value and quick wins throughout the Autonomous car liability project lifecycle?
<--- Score

43. Identify an operational issue in your organization, for example, could a particular task be done more quickly or more efficiently by Autonomous car liability?
<--- Score

44. How do you use Autonomous car liability data and information to support organizational decision making and innovation?
<--- Score

45. How was the detailed process map generated, verified, and validated?
<--- Score

46. What information qualified as important?
<--- Score

47. What type of niche opportunities can be pursued?
<--- Score

48. Are your outputs consistent?
<--- Score

49. What process should you select for improvement?
<--- Score

50. How many input/output points does it require?
<--- Score

51. What are evaluation criteria for the output?
<--- Score

52. Is the gap/opportunity displayed and communicated in financial terms?
<--- Score

53. How will corresponding data be collected?
<--- Score

54. Is there a strict change management process?
<--- Score

55. What resources go in to get the desired output?
<--- Score

56. How is the way you as the leader think and process information affecting your organizational culture?
<--- Score

57. Will the term product include all forms of software and data?
<--- Score

58. Is pre-qualification of suppliers carried out?
<--- Score

59. What, related to, Autonomous car liability processes does your organization outsource?
<--- Score

60. How do your work systems and key work processes relate to and capitalize on your core competencies?
<--- Score

61. How will the data be checked for quality?
<--- Score

62. When should a process be art not science?
<--- Score

63. What is the Value Stream Mapping?
<--- Score

64. What systems/processes must you excel at?
<--- Score

65. How do you promote understanding that opportunity for improvement is not criticism of the status quo, or the people who created the status quo?
<--- Score

66. Who will gather what data?
<--- Score

67. What conclusions were drawn from the team's data collection and analysis? How did the team reach these conclusions?
<--- Score

68. Who owns what data?
<--- Score

69. What Autonomous car liability data do you gather or use now?
<--- Score

70. What data do you need to collect?
<--- Score

71. What output to create?
<--- Score

72. What are your outputs?
<--- Score

73. What Autonomous car liability metrics are outputs of the process?
<--- Score

74. Is the suppliers process defined and controlled?
<--- Score

75. How do you identify specific Autonomous car liability investment opportunities and emerging trends?
<--- Score

76. What other jobs or tasks affect the performance of the steps in the Autonomous car liability process?
<--- Score

77. What are your Autonomous car liability processes?
<--- Score

78. What is your organizations system for selecting qualified vendors?
<--- Score

79. How is Autonomous car liability data gathered?
<--- Score

80. How do you define collaboration and team output?
<--- Score

81. Has an output goal been set?
<--- Score

82. What are the Autonomous car liability design outputs?
<--- Score

83. Do you have the authority to produce the output?
<--- Score

84. How difficult is it to qualify what Autonomous car liability ROI is?
<--- Score

85. How is the Autonomous car liability Value Stream Mapping managed?
<--- Score

86. Were any designed experiments used to generate additional insight into the data analysis?
<--- Score

87. What is the oversight process?
<--- Score

88. What is your organizations process which leads to recognition of value generation?
<--- Score

89. How will the change process be managed?
<--- Score

90. What are the processes for audit reporting and management?
<--- Score

91. Who will facilitate the team and process?
<--- Score

92. What is the output?
<--- Score

93. Did any value-added analysis or 'lean thinking' take place to identify some of the gaps shown on the 'as is' process map?
<--- Score

94. Are you missing Autonomous car liability opportunities?
<--- Score

95. How often will data be collected for measures?
<--- Score

96. What tools were used to generate the list of possible causes?
<--- Score

97. What are the Autonomous car liability business drivers?
<--- Score

98. What internal processes need improvement?
<--- Score

99. Is data and process analysis, root cause analysis and quantifying the gap/opportunity in place?
<--- Score

100. What Autonomous car liability data should be managed?
<--- Score

101. What are your current levels and trends in key measures or indicators of Autonomous car liability product and process performance that are important to and directly serve your customers? How do these results compare with the performance of your competitors and other organizations with similar offerings?
<--- Score

102. What were the financial benefits resulting from any 'ground fruit or low-hanging fruit' (quick fixes)?
<--- Score

103. Is there an established change management process?
<--- Score

104. A compounding model resolution with available relevant data can often provide insight towards a solution methodology; which Autonomous car

liability models, tools and techniques are necessary?
<--- Score

105. Were there any improvement opportunities identified from the process analysis?
<--- Score

106. What qualifications are needed?
<--- Score

107. What process improvements will be needed?
<--- Score

108. Is the final output clearly identified?
<--- Score

109. What are the disruptive Autonomous car liability technologies that enable your organization to radically change your business processes?
<--- Score

110. What controls do you have in place to protect data?
<--- Score

111. What is the Autonomous car liability Driver?
<--- Score

112. Where is Autonomous car liability data gathered?
<--- Score

113. How is the data gathered?
<--- Score

114. Are all staff in core Autonomous car liability subjects Highly Qualified?

<--- Score

115. Should you invest in industry-recognized qualifications?
<--- Score

116. Were Pareto charts (or similar) used to portray the 'heavy hitters' (or key sources of variation)?
<--- Score

117. Have any additional benefits been identified that will result from closing all or most of the gaps?
<--- Score

118. Where is the data coming from to measure compliance?
<--- Score

119. Do quality systems drive continuous improvement?
<--- Score

120. Have the problem and goal statements been updated to reflect the additional knowledge gained from the analyze phase?
<--- Score

121. Do you understand your management processes today?
<--- Score

122. What are your key performance measures or indicators and in-process measures for the control and improvement of your Autonomous car liability processes?
<--- Score

123. What is the cost of poor quality as supported by the team's analysis?
<--- Score

124. What types of data do your Autonomous car liability indicators require?
<--- Score

125. Do several people in different organizational units assist with the Autonomous car liability process?
<--- Score

126. Are gaps between current performance and the goal performance identified?
<--- Score

127. Was a detailed process map created to amplify critical steps of the 'as is' stakeholder process?
<--- Score

128. What are the best opportunities for value improvement?
<--- Score

129. What did the team gain from developing a sub-process map?
<--- Score

130. How can risk management be tied procedurally to process elements?
<--- Score

131. Can you add value to the current Autonomous car liability decision-making process (largely qualitative) by incorporating uncertainty modeling

(more quantitative)?
<--- Score

132. Are Autonomous car liability changes recognized early enough to be approved through the regular process?
<--- Score

133. Where can you get qualified talent today?
<--- Score

134. Has data output been validated?
<--- Score

135. How has the Autonomous car liability data been gathered?
<--- Score

Add up total points for this section:
_____ = Total points for this section

Divided by: ______ (number of statements answered) = ______
Average score for this section

Transfer your score to the Autonomous car liability Index at the beginning of the Self-Assessment.

CRITERION #5: IMPROVE:

INTENT: Develop a practical solution. Innovate, establish and test the solution and to measure the results.

In my belief, the answer to this question is clearly defined:

5 Strongly Agree

4 Agree

3 Neutral

2 Disagree

1 Strongly Disagree

1. What area needs the greatest improvement?
<--- Score

2. Who do you report Autonomous car liability results to?
<--- Score

3. How do you go about comparing Autonomous car liability approaches/solutions?

<--- Score

4. Have you identified breakpoints and/or risk tolerances that will trigger broad consideration of a potential need for intervention or modification of strategy?
<--- Score

5. Is Autonomous car liability documentation maintained?
<--- Score

6. How can you improve performance?
<--- Score

7. What error proofing will be done to address some of the discrepancies observed in the 'as is' process?
<--- Score

8. Who should make the Autonomous car liability decisions?
<--- Score

9. How do you link measurement and risk?
<--- Score

10. What strategies for Autonomous car liability improvement are successful?
<--- Score

11. What tools were used to evaluate the potential solutions?
<--- Score

12. Are you assessing Autonomous car liability and risk?

<--- Score

13. Risk events: what are the things that could go wrong?
<--- Score

14. Do vendor agreements bring new compliance risk ?
<--- Score

15. How will you recognize and celebrate results?
<--- Score

16. How are Autonomous car liability risks managed?
<--- Score

17. What is the magnitude of the improvements?
<--- Score

18. What are the appropriate crash metrics for evaluating the relative safety performance of an autonomous vehicle?
<--- Score

19. Who manages Autonomous car liability risk?
<--- Score

20. What are the concrete Autonomous car liability results?
<--- Score

21. Who controls key decisions that will be made?
<--- Score

22. Are risk triggers captured?
<--- Score

23. Does the goal represent a desired result that can be measured?
<--- Score

24. Is there a high likelihood that any recommendations will achieve their intended results?
<--- Score

25. Where do you need Autonomous car liability improvement?
<--- Score

26. What attendant changes will need to be made to ensure that the solution is successful?
<--- Score

27. What is the implementation plan?
<--- Score

28. How do you measure risk?
<--- Score

29. Was a Autonomous car liability charter developed?
<--- Score

30. What alternative responses are available to manage risk?
<--- Score

31. How will you measure the results?
<--- Score

32. Where do the Autonomous car liability decisions reside?
<--- Score

33. Are the key business and technology risks being managed?
<--- Score

34. How scalable is your Autonomous car liability solution?
<--- Score

35. How risky is your organization?
<--- Score

36. Risk factors: what are the characteristics of Autonomous car liability that make it risky?
<--- Score

37. Who are the Autonomous car liability decision-makers?
<--- Score

38. How is continuous improvement applied to risk management?
<--- Score

39. Can you integrate quality management and risk management?
<--- Score

40. What were the criteria for evaluating a Autonomous car liability pilot?
<--- Score

41. What is Autonomous car liability's impact on utilizing the best solution(s)?
<--- Score

42. Is there any other Autonomous car liability solution?
<--- Score

43. How can you better manage risk?
<--- Score

44. How do you mitigate Autonomous car liability risk?
<--- Score

45. What assumptions are made about the solution and approach?
<--- Score

46. How are policy decisions made and where?
<--- Score

47. To what extent does management recognize Autonomous car liability as a tool to increase the results?
<--- Score

48. How do you improve your likelihood of success ?
<--- Score

49. Risk Identification: What are the possible risk events your organization faces in relation to Autonomous car liability?
<--- Score

50. What tools were used to tap into the creativity and encourage 'outside the box' thinking?
<--- Score

51. How do you define the solutions' scope?

<--- Score

52. What Autonomous car liability improvements can be made?
<--- Score

53. Do those selected for the Autonomous car liability team have a good general understanding of what Autonomous car liability is all about?
<--- Score

54. How do you measure progress and evaluate training effectiveness?
<--- Score

55. What needs improvement? Why?
<--- Score

56. Are decisions made in a timely manner?
<--- Score

57. Is the Autonomous car liability solution sustainable?
<--- Score

58. What does the 'should be' process map/design look like?
<--- Score

59. In the past few months, what is the smallest change you have made that has had the biggest positive result? What was it about that small change that produced the large return?
<--- Score

60. What actually has to improve and by how much?

<--- Score

61. How is knowledge sharing about risk management improved?
<--- Score

62. What criteria will you use to assess your Autonomous car liability risks?
<--- Score

63. Do you have the optimal project management team structure?
<--- Score

64. What are the implications of the one critical Autonomous car liability decision 10 minutes, 10 months, and 10 years from now?
<--- Score

65. Is risk periodically assessed?
<--- Score

66. Is the Autonomous car liability risk managed?
<--- Score

67. How do you deal with Autonomous car liability risk?
<--- Score

68. When will customers begin to demand discounts given a reduced risk profile?
<--- Score

69. How can skill-level changes improve Autonomous car liability?
<--- Score

70. Who are the key stakeholders for the Autonomous car liability evaluation?
<--- Score

71. What current systems have to be understood and/ or changed?
<--- Score

72. Who makes the Autonomous car liability decisions in your organization?
<--- Score

73. How do the Autonomous car liability results compare with the performance of your competitors and other organizations with similar offerings?
<--- Score

74. Why improve in the first place?
<--- Score

75. How can you improve Autonomous car liability?
<--- Score

76. What are the affordable Autonomous car liability risks?
<--- Score

77. Have you achieved Autonomous car liability improvements?
<--- Score

78. Are risk management tasks balanced centrally and locally?
<--- Score

79. Would you develop a Autonomous car liability Communication Strategy?
<--- Score

80. Is the solution technically practical?
<--- Score

81. How will you know when its improved?
<--- Score

82. Who are the people involved in developing and implementing Autonomous car liability?
<--- Score

83. What risks do you need to manage?
<--- Score

84. How will you know that you have improved?
<--- Score

85. Who will be responsible for documenting the Autonomous car liability requirements in detail?
<--- Score

86. Are events managed to resolution?
<--- Score

87. Who are the Autonomous car liability decision makers?
<--- Score

88. What can you do to improve?
<--- Score

89. Are the risks fully understood, reasonable and manageable?

<--- Score

90. Who manages supplier risk management in your organization?
<--- Score

91. How do you keep improving Autonomous car liability?
<--- Score

92. Is supporting Autonomous car liability documentation required?
<--- Score

93. For decision problems, how do you develop a decision statement?
<--- Score

94. How does your organization evaluate strategic Autonomous car liability success?
<--- Score

95. What should a proof of concept or pilot accomplish?
<--- Score

96. Which of the recognised risks out of all risks can be most likely transferred?
<--- Score

97. Autonomous car liability risk decisions: whose call Is It?
<--- Score

98. How do you manage Autonomous car liability risk?
<--- Score

99. What is the Autonomous car liability's sustainability risk?
<--- Score

100. When you map the key players in your own work and the types/domains of relationships with them, which relationships do you find easy and which challenging, and why?
<--- Score

101. How do you manage and improve your Autonomous car liability work systems to deliver customer value and achieve organizational success and sustainability?
<--- Score

102. What resources are required for the improvement efforts?
<--- Score

103. What to do with the results or outcomes of measurements?
<--- Score

104. What do you want to improve?
<--- Score

105. What are the Autonomous car liability security risks?
<--- Score

106. What tools were most useful during the improve phase?
<--- Score

107. What is the relevant baseline for evaluating the relative safety of an autonomous vehicle?
<--- Score

108. Is the scope clearly documented?
<--- Score

109. Can you identify any significant risks or exposures to Autonomous car liability third- parties (vendors, service providers, alliance partners etc) that concern you?
<--- Score

110. How do you improve Autonomous car liability service perception, and satisfaction?
<--- Score

111. How does the team improve its work?
<--- Score

112. At what point will vulnerability assessments be performed once Autonomous car liability is put into production (e.g., ongoing Risk Management after implementation)?
<--- Score

113. Which Autonomous car liability solution is appropriate?
<--- Score

114. Are the most efficient solutions problem-specific?
<--- Score

115. For estimation problems, how do you develop an estimation statement?
<--- Score

116. Explorations of the frontiers of Autonomous car liability will help you build influence, improve Autonomous car liability, optimize decision making, and sustain change, what is your approach?
<--- Score

117. What tools do you use once you have decided on a Autonomous car liability strategy and more importantly how do you choose?
<--- Score

118. How do you improve productivity?
<--- Score

119. Who will be responsible for making the decisions to include or exclude requested changes once Autonomous car liability is underway?
<--- Score

120. How do you decide how much to remunerate an employee?
<--- Score

121. What are the expected Autonomous car liability results?
<--- Score

122. Can the solution be designed and implemented within an acceptable time period?
<--- Score

123. What communications are necessary to support the implementation of the solution?
<--- Score

124. Is any Autonomous car liability documentation required?
<--- Score

125. Do you need to do a usability evaluation?
<--- Score

126. What practices helps your organization to develop its capacity to recognize patterns?
<--- Score

127. Are procedures documented for managing Autonomous car liability risks?
<--- Score

128. What is the team's contingency plan for potential problems occurring in implementation?
<--- Score

129. Do you cover the five essential competencies: Communication, Collaboration,Innovation, Adaptability, and Leadership that improve an organizations ability to leverage the new Autonomous car liability in a volatile global economy?
<--- Score

130. What improvements have been achieved?
<--- Score

131. Do you combine technical expertise with business knowledge and Autonomous car liability Key topics include lifecycles, development approaches, requirements and how to make a business case?
<--- Score

132. What were the underlying assumptions on the

cost-benefit analysis?
<--- Score

133. What lessons, if any, from a pilot were incorporated into the design of the full-scale solution?
<--- Score

134. What are your current levels and trends in key measures or indicators of workforce and leader development?
<--- Score

135. How will you know that a change is an improvement?
<--- Score

136. What is Autonomous car liability risk?
<--- Score

137. How significant is the improvement in the eyes of the end user?
<--- Score

138. Who will be using the results of the measurement activities?
<--- Score

139. Who controls the risk?
<--- Score

Add up total points for this section:
_____ = Total points for this section

Divided by: ______ (number of statements answered) = ______
Average score for this section

Transfer your score to the Autonomous car liability Index at the beginning of the Self-Assessment.

CRITERION #6: CONTROL:

INTENT: Implement the practical solution. Maintain the performance and correct possible complications.

In my belief, the answer to this question is clearly defined:

5 Strongly Agree

4 Agree

3 Neutral

2 Disagree

1 Strongly Disagree

1. Is there a standardized process?
<--- Score

2. Can support from partners be adjusted?
<--- Score

3. How likely is the current Autonomous car liability plan to come in on schedule or on budget?
<--- Score

4. What adjustments to the strategies are needed?
<--- Score

5. Are the planned controls in place?
<--- Score

6. In the case of a Autonomous car liability project, the criteria for the audit derive from implementation objectives, an audit of a Autonomous car liability project involves assessing whether the recommendations outlined for implementation have been met, can you track that any Autonomous car liability project is implemented as planned, and is it working?
<--- Score

7. Does job training on the documented procedures need to be part of the process team's education and training?
<--- Score

8. Can you adapt and adjust to changing Autonomous car liability situations?
<--- Score

9. Who is going to spread your message?
<--- Score

10. Are there documented procedures?
<--- Score

11. What Autonomous car liability standards are applicable?
<--- Score

12. How will report readings be checked to effectively monitor performance?
<--- Score

13. Will the team be available to assist members in planning investigations?
<--- Score

14. What can you control?
<--- Score

15. Has the improved process and its steps been standardized?
<--- Score

16. How do you plan on providing proper recognition and disclosure of supporting companies?
<--- Score

17. How will Autonomous car liability decisions be made and monitored?
<--- Score

18. How widespread is its use?
<--- Score

19. Is a response plan in place for when the input, process, or output measures indicate an 'out-of-control' condition?
<--- Score

20. How do senior leaders actions reflect a commitment to the organizations Autonomous car liability values?
<--- Score

21. What should the next improvement project be that is related to Autonomous car liability?
<--- Score

22. Has the Autonomous car liability value of standards been quantified?
<--- Score

23. What quality tools were useful in the control phase?
<--- Score

24. Does Autonomous car liability appropriately measure and monitor risk?
<--- Score

25. What key inputs and outputs are being measured on an ongoing basis?
<--- Score

26. How will the process owner verify improvement in present and future sigma levels, process capabilities?
<--- Score

27. Who controls critical resources?
<--- Score

28. What do you stand for--and what are you against?
<--- Score

29. Will your goals reflect your program budget?
<--- Score

30. What other areas of the group might benefit from the Autonomous car liability team's improvements, knowledge, and learning?

<--- Score

31. Are new process steps, standards, and documentation ingrained into normal operations?
<--- Score

32. Do the viable solutions scale to future needs?
<--- Score

33. What should you measure to verify efficiency gains?
<--- Score

34. What is your plan to assess your security risks?
<--- Score

35. Is the Autonomous car liability test/monitoring cost justified?
<--- Score

36. Is reporting being used or needed?
<--- Score

37. How do your controls stack up?
<--- Score

38. Are pertinent alerts monitored, analyzed and distributed to appropriate personnel?
<--- Score

39. How do you plan for the cost of succession?
<--- Score

40. How do controls support value?
<--- Score

41. What are your results for key measures or indicators of the accomplishment of your Autonomous car liability strategy and action plans, including building and strengthening core competencies?
<--- Score

42. How will new or emerging customer needs/ requirements be checked/communicated to orient the process toward meeting the new specifications and continually reducing variation?
<--- Score

43. Against what alternative is success being measured?
<--- Score

44. Is there a documented and implemented monitoring plan?
<--- Score

45. Are the Autonomous car liability standards challenging?
<--- Score

46. Who is the Autonomous car liability process owner?
<--- Score

47. What are you attempting to measure/monitor?
<--- Score

48. Do the Autonomous car liability decisions you make today help people and the planet tomorrow?
<--- Score

49. Are controls in place and consistently applied?
<--- Score

50. Have new or revised work instructions resulted?
<--- Score

51. Are you measuring, monitoring and predicting Autonomous car liability activities to optimize operations and profitability, and enhancing outcomes?
<--- Score

52. How do you spread information?
<--- Score

53. How will input, process, and output variables be checked to detect for sub-optimal conditions?
<--- Score

54. What is the best design framework for Autonomous car liability organization now that, in a post industrial-age if the top-down, command and control model is no longer relevant?
<--- Score

55. What is the recommended frequency of auditing?
<--- Score

56. Is knowledge gained on process shared and institutionalized?
<--- Score

57. Is a response plan established and deployed?
<--- Score

58. Is there a recommended audit plan for routine

surveillance inspections of Autonomous car liability's gains?
<--- Score

59. Are suggested corrective/restorative actions indicated on the response plan for known causes to problems that might surface?
<--- Score

60. Implementation Planning: is a pilot needed to test the changes before a full roll out occurs?
<--- Score

61. Are operating procedures consistent?
<--- Score

62. How might the group capture best practices and lessons learned so as to leverage improvements?
<--- Score

63. How will you measure your QA plan's effectiveness?
<--- Score

64. What are the known security controls?
<--- Score

65. How can you best use all of your knowledge repositories to enhance learning and sharing?
<--- Score

66. Will existing staff require re-training, for example, to learn new business processes?
<--- Score

67. Where do ideas that reach policy makers and

planners as proposals for Autonomous car liability strengthening and reform actually originate?
<--- Score

68. What are the key elements of your Autonomous car liability performance improvement system, including your evaluation, organizational learning, and innovation processes?
<--- Score

69. What do you measure to verify effectiveness gains?
<--- Score

70. Do you monitor the Autonomous car liability decisions made and fine tune them as they evolve?
<--- Score

71. How do you select, collect, align, and integrate Autonomous car liability data and information for tracking daily operations and overall organizational performance, including progress relative to strategic objectives and action plans?
<--- Score

72. Is there a transfer of ownership and knowledge to process owner and process team tasked with the responsibilities.
<--- Score

73. Who has control over resources?
<--- Score

74. What are the performance and scale of the Autonomous car liability tools?
<--- Score

75. Does the Autonomous car liability performance meet the customer's requirements?
<--- Score

76. What is the control/monitoring plan?
<--- Score

77. How do you monitor usage and cost?
<--- Score

78. Act/Adjust: What Do you Need to Do Differently?
<--- Score

79. Is new knowledge gained imbedded in the response plan?
<--- Score

80. Do you monitor the effectiveness of your Autonomous car liability activities?
<--- Score

81. How is Autonomous car liability project cost planned, managed, monitored?
<--- Score

82. How is change control managed?
<--- Score

83. How will the day-to-day responsibilities for monitoring and continual improvement be transferred from the improvement team to the process owner?
<--- Score

84. What other systems, operations, processes, and

infrastructures (hiring practices, staffing, training, incentives/rewards, metrics/dashboards/scorecards, etc.) need updates, additions, changes, or deletions in order to facilitate knowledge transfer and improvements?
<--- Score

85. Are the planned controls working?
<--- Score

86. Is there documentation that will support the successful operation of the improvement?
<--- Score

87. What are the critical parameters to watch?
<--- Score

88. Is there an action plan in case of emergencies?
<--- Score

89. Will any special training be provided for results interpretation?
<--- Score

90. How do you encourage people to take control and responsibility?
<--- Score

91. How will the process owner and team be able to hold the gains?
<--- Score

92. Who sets the Autonomous car liability standards?
<--- Score

93. What do your reports reflect?

<--- Score

94. How do you establish and deploy modified action plans if circumstances require a shift in plans and rapid execution of new plans?
<--- Score

95. Who will be in control?
<--- Score

96. Are documented procedures clear and easy to follow for the operators?
<--- Score

97. Does the response plan contain a definite closed loop continual improvement scheme (e.g., plan-do-check-act)?
<--- Score

98. Is there a Autonomous car liability Communication plan covering who needs to get what information when?
<--- Score

99. Is there a control plan in place for sustaining improvements (short and long-term)?
<--- Score

100. Does a troubleshooting guide exist or is it needed?
<--- Score

Add up total points for this section:
_____ = Total points for this section

Divided by: ______ (number of

statements answered) = ______
Average score for this section

Transfer your score to the Autonomous car liability Index at the beginning of the Self-Assessment.

CRITERION #7: SUSTAIN:

INTENT: Retain the benefits.

In my belief, the answer to this question is clearly defined:

5 Strongly Agree

4 Agree

3 Neutral

2 Disagree

1 Strongly Disagree

1. What happens when a new employee joins the organization?
<--- Score

2. Is maximizing Autonomous car liability protection the same as minimizing Autonomous car liability loss?
<--- Score

3. What trouble can you get into?
<--- Score

4. How much does Autonomous car liability help?
<--- Score

5. What threat is Autonomous car liability addressing?
<--- Score

6. What must you excel at?
<--- Score

7. How can you become more high-tech but still be high touch?
<--- Score

8. Do you think Autonomous car liability accomplishes the goals you expect it to accomplish?
<--- Score

9. Do you think you know, or do you know you know ?
<--- Score

10. What is the recommended frequency of auditing?
<--- Score

11. How can you incorporate support to ensure safe and effective use of Autonomous car liability into the services that you provide?
<--- Score

12. How do you provide a safe environment -physically and emotionally?
<--- Score

13. How do you keep records, of what?
<--- Score

14. Has implementation been effective in reaching

specified objectives so far?
<--- Score

15. What is your question? Why?
<--- Score

16. Do you know what you are doing? And who do you call if you don't?
<--- Score

17. How do you maintain Autonomous car liability's Integrity?
<--- Score

18. How do you stay inspired?
<--- Score

19. What is it like to work for you?
<--- Score

20. What are specific Autonomous car liability rules to follow?
<--- Score

21. How do you listen to customers to obtain actionable information?
<--- Score

22. What could happen if you do not do it?
<--- Score

23. Have benefits been optimized with all key stakeholders?
<--- Score

24. Who will determine interim and final deadlines?

<--- Score

25. How do you manage Autonomous car liability Knowledge Management (KM)?
<--- Score

26. Do you know how protection system from your device works?
<--- Score

27. Are the criteria for selecting recommendations stated?
<--- Score

28. Is your strategy driving your strategy? Or is the way in which you allocate resources driving your strategy?
<--- Score

29. How do you create buy-in?
<--- Score

30. Do you know who is a friend or a foe?
<--- Score

31. Who, on the executive team or the board, has spoken to a customer recently?
<--- Score

32. Who are the key stakeholders?
<--- Score

33. What are the key enablers to make this Autonomous car liability move?
<--- Score

34. What was the last experiment you ran?
<--- Score

35. Is the impact that Autonomous car liability has shown?
<--- Score

36. What would you recommend your friend do if he/ she were facing this dilemma?
<--- Score

37. Who are your customers?
<--- Score

38. At what moment would you think; Will I get fired?
<--- Score

39. Whom among your colleagues do you trust, and for what?
<--- Score

40. How do you know if you are successful?
<--- Score

41. If you had to rebuild your organization without any traditional competitive advantages (i.e., no killer technology, promising research, innovative product/ service delivery model, etcetera), how would your people have to approach their work and collaborate together in order to create the necessary conditions for success?
<--- Score

42. How will you know that the Autonomous car liability project has been successful?
<--- Score

43. How do you go about securing Autonomous car liability?
<--- Score

44. Are there any activities that you can take off your to do list?
<--- Score

45. To whom do you add value?
<--- Score

46. How much contingency will be available in the budget?
<--- Score

47. Why do and why don't your customers like your organization?
<--- Score

48. What is a feasible sequencing of reform initiatives over time?
<--- Score

49. How do you foster the skills, knowledge, talents, attributes, and characteristics you want to have?
<--- Score

50. Is a Autonomous car liability team work effort in place?
<--- Score

51. What did you miss in the interview for the worst hire you ever made?
<--- Score

52. Why not do Autonomous car liability?
<--- Score

53. What Autonomous car liability skills are most important?
<--- Score

54. What goals did you miss?
<--- Score

55. In the past year, what have you done (or could you have done) to increase the accurate perception of your company/brand as ethical and honest?
<--- Score

56. What does your signature ensure?
<--- Score

57. Who is responsible for ensuring appropriate resources (time, people and money) are allocated to Autonomous car liability?
<--- Score

58. Is Autonomous car liability realistic, or are you setting yourself up for failure?
<--- Score

59. What have been your experiences in defining long range Autonomous car liability goals?
<--- Score

60. How long will it take to change?
<--- Score

61. Is the Autonomous car liability organization completing tasks effectively and efficiently?

<--- Score

62. What unique value proposition (UVP) do you offer?
<--- Score

63. Were lessons learned captured and communicated?
<--- Score

64. How will travel demand affect energy and emissions?
<--- Score

65. What is the overall talent health of your organization as a whole at senior levels, and for each organization reporting to a member of the Senior Leadership Team?
<--- Score

66. How do you assess the Autonomous car liability pitfalls that are inherent in implementing it?
<--- Score

67. What are the essentials of internal Autonomous car liability management?
<--- Score

68. What do we do when new problems arise?
<--- Score

69. Do you have enough freaky customers in your portfolio pushing you to the limit day in and day out?
<--- Score

70. Who will provide the final approval of Autonomous car liability deliverables?

<--- Score

71. What are you trying to prove to yourself, and how might it be hijacking your life and business success?
<--- Score

72. Did your employees make progress today?
<--- Score

73. Are you paying enough attention to the partners your company depends on to succeed?
<--- Score

74. Do you have past Autonomous car liability successes?
<--- Score

75. How do you determine the key elements that affect Autonomous car liability workforce satisfaction, how are these elements determined for different workforce groups and segments?
<--- Score

76. If there were zero limitations, what would you do differently?
<--- Score

77. Ask yourself: how would you do this work if you only had one staff member to do it?
<--- Score

78. How do you set Autonomous car liability stretch targets and how do you get people to not only participate in setting these stretch targets but also that they strive to achieve these?
<--- Score

79. Why will customers want to buy your organizations products/services?
<--- Score

80. In a project to restructure Autonomous car liability outcomes, which stakeholders would you involve?
<--- Score

81. Whose voice (department, ethnic group, women, older workers, etc) might you have missed hearing from in your company, and how might you amplify this voice to create positive momentum for your business?
<--- Score

82. How will your organization educate its employees?
<--- Score

83. What are internal and external Autonomous car liability relations?
<--- Score

84. How do you cross-sell and up-sell your Autonomous car liability success?
<--- Score

85. What are the potential basics of Autonomous car liability fraud?
<--- Score

86. Operational - will it work?
<--- Score

87. Is there any existing Autonomous car liability

governance structure?
<--- Score

88. How do you make it meaningful in connecting Autonomous car liability with what users do day-to-day?
<--- Score

89. Are new benefits received and understood?
<--- Score

90. What is your Autonomous car liability strategy?
<--- Score

91. How likely is it that a customer would recommend your company to a friend or colleague?
<--- Score

92. Are you relevant? Will you be relevant five years from now? Ten?
<--- Score

93. In retrospect, of the projects that you pulled the plug on, what percent do you wish had been allowed to keep going, and what percent do you wish had ended earlier?
<--- Score

94. What are the challenges?
<--- Score

95. What are the long-term Autonomous car liability goals?
<--- Score

96. What should you stop doing?

<--- Score

97. What is the range of capabilities?
<--- Score

98. What have you done to protect your business from competitive encroachment?
<--- Score

99. Are assumptions made in Autonomous car liability stated explicitly?
<--- Score

100. Why is it important to have senior management support for a Autonomous car liability project?
<--- Score

101. What counts that you are not counting?
<--- Score

102. Who will manage the integration of tools?
<--- Score

103. What are your personal philosophies regarding Autonomous car liability and how do they influence your work?
<--- Score

104. Which Autonomous car liability goals are the most important?
<--- Score

105. What is effective Autonomous car liability?
<--- Score

106. Marketing budgets are tighter, consumers are

more skeptical, and social media has changed forever the way we talk about Autonomous car liability, how do you gain traction?
<--- Score

107. How do you ensure that implementations of Autonomous car liability products are done in a way that ensures safety?
<--- Score

108. What is the overall business strategy?
<--- Score

109. Are you / should you be revolutionary or evolutionary?
<--- Score

110. What is the funding source for this project?
<--- Score

111. Which models, tools and techniques are necessary?
<--- Score

112. Who do you think the world wants your organization to be?
<--- Score

113. What are the business goals Autonomous car liability is aiming to achieve?
<--- Score

114. What may be the consequences for the performance of an organization if all stakeholders are not consulted regarding Autonomous car liability?
<--- Score

115. Think of your Autonomous car liability project, what are the main functions?
<--- Score

116. How can you become the company that would put you out of business?
<--- Score

117. If you find that you havent accomplished one of the goals for one of the steps of the Autonomous car liability strategy, what will you do to fix it?
<--- Score

118. Why is Autonomous car liability important for you now?
<--- Score

119. What are strategies for increasing support and reducing opposition?
<--- Score

120. Is there a work around that you can use?
<--- Score

121. Would you rather sell to knowledgeable and informed customers or to uninformed customers?
<--- Score

122. Who else should you help?
<--- Score

123. How will you insure seamless interoperability of Autonomous car liability moving forward?
<--- Score

124. Are the assumptions believable and achievable?
<--- Score

125. Do you feel that more should be done in the Autonomous car liability area?
<--- Score

126. What is something you believe that nearly no one agrees with you on?
<--- Score

127. Is a Autonomous car liability breakthrough on the horizon?
<--- Score

128. What are the barriers to increased Autonomous car liability production?
<--- Score

129. If you were responsible for initiating and implementing major changes in your organization, what steps might you take to ensure acceptance of those changes?
<--- Score

130. Are you satisfied with your current role? If not, what is missing from it?
<--- Score

131. Do Autonomous car liability rules make a reasonable demand on a users capabilities?
<--- Score

132. What one word do you want to own in the minds of your customers, employees, and partners?
<--- Score

133. What you are going to do to affect the numbers?
<--- Score

134. Will it be accepted by users?
<--- Score

135. What is an unauthorized commitment?
<--- Score

136. Who is responsible for errors?
<--- Score

137. Who is on the team?
<--- Score

138. How will you motivate the stakeholders with the least vested interest?
<--- Score

139. Who do you want your customers to become?
<--- Score

140. What management system can you use to leverage the Autonomous car liability experience, ideas, and concerns of the people closest to the work to be done?
<--- Score

141. What is the kind of project structure that would be appropriate for your Autonomous car liability project, should it be formal and complex, or can it be less formal and relatively simple?
<--- Score

142. What happens if you do not have enough

funding?
<--- Score

143. How do you deal with Autonomous car liability changes?
<--- Score

144. How do you track customer value, profitability or financial return, organizational success, and sustainability?
<--- Score

145. Do you have an implicit bias for capital investments over people investments?
<--- Score

146. What Autonomous car liability modifications can you make work for you?
<--- Score

147. How do you accomplish your long range Autonomous car liability goals?
<--- Score

148. Who is responsible for Autonomous car liability?
<--- Score

149. What is the estimated value of the project?
<--- Score

150. What will be the consequences to the stakeholder (financial, reputation etc) if Autonomous car liability does not go ahead or fails to deliver the objectives?
<--- Score

151. How do you proactively clarify deliverables and Autonomous car liability quality expectations?
<--- Score

152. What are the rules and assumptions your industry operates under? What if the opposite were true?
<--- Score

153. What is your BATNA (best alternative to a negotiated agreement)?
<--- Score

154. What is the craziest thing you can do?
<--- Score

155. Who are four people whose careers you have enhanced?
<--- Score

156. What happens at your organization when people fail?
<--- Score

157. What is the source of the strategies for Autonomous car liability strengthening and reform?
<--- Score

158. If no one would ever find out about your accomplishments, how would you lead differently?
<--- Score

159. What are your most important goals for the strategic Autonomous car liability objectives?
<--- Score

160. Do you have the right people on the bus?

<--- Score

161. Who uses your product in ways you never expected?
<--- Score

162. What are the top 3 things at the forefront of your Autonomous car liability agendas for the next 3 years?
<--- Score

163. How do you keep the momentum going?
<--- Score

164. Are you using a design thinking approach and integrating Innovation, Autonomous car liability Experience, and Brand Value?
<--- Score

165. Will there be any necessary staff changes (redundancies or new hires)?
<--- Score

166. Are you maintaining a past–present–future perspective throughout the Autonomous car liability discussion?
<--- Score

167. Is there any reason to believe the opposite of my current belief?
<--- Score

168. What information is critical to your organization that your executives are ignoring?
<--- Score

169. Is Autonomous car liability dependent on the

successful delivery of a current project?
<--- Score

170. Do you have the right capabilities and capacities?
<--- Score

171. How do you engage the workforce, in addition to satisfying them?
<--- Score

172. Where can you break convention?
<--- Score

173. How do customers see your organization?
<--- Score

174. What would have to be true for the option on the table to be the best possible choice?
<--- Score

175. If you got fired and a new hire took your place, what would she do different?
<--- Score

176. What are current Autonomous car liability paradigms?
<--- Score

177. What are the short and long-term Autonomous car liability goals?
<--- Score

178. What is your formula for success in Autonomous car liability ?
<--- Score

179. Can you maintain your growth without detracting from the factors that have contributed to your success?
<--- Score

180. Who will be responsible for deciding whether Autonomous car liability goes ahead or not after the initial investigations?
<--- Score

181. How do you lead with Autonomous car liability in mind?
<--- Score

182. How are you doing compared to your industry?
<--- Score

183. What business benefits will Autonomous car liability goals deliver if achieved?
<--- Score

184. Do you say no to customers for no reason?
<--- Score

185. How does Autonomous car liability integrate with other stakeholder initiatives?
<--- Score

186. How much should the manufacturer charge for its expected liabilities?
<--- Score

187. If you weren't already in this business, would you enter it today? And if not, what are you going to do about it?
<--- Score

188. Why should people listen to you?
<--- Score

189. What is your competitive advantage?
<--- Score

190. What are you challenging?
<--- Score

191. How do you govern and fulfill your societal responsibilities?
<--- Score

192. Who do we want your customers to become?
<--- Score

193. What stupid rule would you most like to kill?
<--- Score

194. What relationships among Autonomous car liability trends do you perceive?
<--- Score

195. What is the big Autonomous car liability idea?
<--- Score

196. What is the purpose of Autonomous car liability in relation to the mission?
<--- Score

197. Which individuals, teams or departments will be involved in Autonomous car liability?
<--- Score

198. Can the schedule be done in the given time?

<--- Score

199. How is implementation research currently incorporated into each of your goals?
<--- Score

200. How important is Autonomous car liability to the user organizations mission?
<--- Score

201. Who have you, as a company, historically been when you've been at your best?
<--- Score

202. What are the gaps in your knowledge and experience?
<--- Score

203. Have new benefits been realized?
<--- Score

204. Why should you adopt a Autonomous car liability framework?
<--- Score

205. Which functions and people interact with the supplier and or customer?
<--- Score

206. If your customer were your grandmother, would you tell her to buy what you're selling?
<--- Score

207. What role does communication play in the success or failure of a Autonomous car liability project?

<--- Score

208. What projects are going on in the organization today, and what resources are those projects using from the resource pools?
<--- Score

209. When information truly is ubiquitous, when reach and connectivity are completely global, when computing resources are infinite, and when a whole new set of impossibilities are not only possible, but happening, what will that do to your business?
<--- Score

210. How would your organizations stakeholders feel about it?
<--- Score

211. How can you negotiate Autonomous car liability successfully with a stubborn boss, an irate client, or a deceitful coworker?
<--- Score

212. Are all key stakeholders present at all Structured Walkthroughs?
<--- Score

213. Political -is anyone trying to undermine this project?
<--- Score

214. Are you changing as fast as the world around you?
<--- Score

Add up total points for this section:

_____ = Total points for this section

Divided by: ______ (number of statements answered) = ______ Average score for this section

Transfer your score to the Autonomous car liability Index at the beginning of the Self-Assessment.

Autonomous car liability and Managing Projects, Criteria for Project Managers:

1.0 Initiating Process Group: Autonomous car liability

1. At which stage, in a typical Autonomous car liability project do stake holders have maximum influence?

2. What are the required resources?

3. Will the Autonomous car liability project meet the client requirements, and will it achieve the business success criteria that justified doing the Autonomous car liability project in the first place?

4. The Autonomous car liability project managers have maximum authority in which type of organization?

5. How will you know you did it?

6. Although the Autonomous car liability project manager does not directly manage procurement and contracting activities, who does manage procurement and contracting activities in your organization then if not the PM?

7. Just how important is your work to the overall success of the Autonomous car liability project?

8. Where must it be done?

9. How should needs be met?

10. The Autonomous car liability project you are managing has nine stakeholders. How many channel of communications are there between corresponding

stakeholders?

11. Mitigate. what will you do to minimize the impact should the risk event occur?

12. Are the Autonomous car liability project team and stakeholders meeting regularly and using a meeting agenda and taking notes to accurately document what is being covered and what happened in the weekly meetings?

13. What were things that you did very well and want to do the same again on the next Autonomous car liability project?

14. Measurable - are the targets measurable?

15. Do you understand the quality and control criteria that must be achieved for successful Autonomous car liability project completion?

16. Have you evaluated the teams performance and asked for feedback?

17. How well did the chosen processes fit the needs of the Autonomous car liability project?

18. Who is performing the work of the Autonomous car liability project?

19. Are you just doing busywork to pass the time?

20. How will it affect me?

1.1 Project Charter: Autonomous car liability

21. Review the general mission What system will be affected by the improvement efforts?

22. What material?

23. What outcome, in measureable terms, are you hoping to accomplish?

24. Why use a Autonomous car liability project charter?

25. What are the known stakeholder requirements?

26. What changes can you make to improve?

27. Success determination factors: how will the success of the Autonomous car liability project be determined from the customers perspective?

28. How will you know a change is an improvement?

29. Will this replace an existing product?

30. Does the Autonomous car liability project need to consider any special capacity or capability issues?

31. What does it need to do?

32. What is the most common tool for helping define the detail?

33. Who ise input and support will this Autonomous car liability project require?

34. What is the justification?

35. How high should you set your goals?

36. What are some examples of a business case?

37. Autonomous car liability project deliverables: what is the Autonomous car liability project going to produce?

38. If finished, on what date did it finish?

39. Pop quiz – which are the same inputs as in the Autonomous car liability project charter?

40. Where does all this information come from?

1.2 Stakeholder Register: Autonomous car liability

41. Who are the stakeholders?

42. How big is the gap?

43. Who is managing stakeholder engagement?

44. What is the power of the stakeholder?

45. Is your organization ready for change?

46. What are the major Autonomous car liability project milestones requiring communications or providing communications opportunities?

47. How should employers make voices heard?

48. How much influence do they have on the Autonomous car liability project?

49. What opportunities exist to provide communications?

50. Who wants to talk about Security?

51. What & Why?

52. How will reports be created?

1.3 Stakeholder Analysis Matrix: Autonomous car liability

53. Arena: in what fields are the actors active, where are they present?

54. What are the mechanisms of public and social accountability, and how can they be made better?

55. What should thwe organizations stakeholders avoid?

56. Resources, assets, people?

57. Where are mitigation costs factored in?

58. What do you need to appraise?

59. Does your organization have bad debt or cash-flow problems?

60. Location and geographical?

61. New technologies, services, ideas?

62. Price, value, quality?

63. How to measure the achievement of the Development Objective?

64. What is the stakeholders name, what is function?

65. What is in it for you?

66. Reliability of data, plan predictability?

67. Morale, commitment, leadership?

68. Market developments?

69. Vital contracts and partners?

70. Timescales, deadlines and pressures?

71. Will the impacts be local, national or international?

72. Who is directly responsible for decisions on issues important to the Autonomous car liability project?

2.0 Planning Process Group: Autonomous car liability

73. Professionals want to know what is expected from them; what are the deliverables?

74. To what extent has a PMO contributed to raising the quality of the design of the Autonomous car liability project?

75. Are work methodologies, financial instruments, etc. shared among departments, organizations and Autonomous car liability projects?

76. How well do the team follow the chosen processes?

77. If action is called for, what form should it take?

78. In what way has the Autonomous car liability project come up with innovative measures for problem-solving?

79. Have more efficient (sensitive) and appropriate measures been adopted to respond to the political and socio-cultural problems identified?

80. If a task is partitionable, is this a sufficient condition to reduce the Autonomous car liability project duration?

81. Is the schedule for the set products being met?

82. What are the different approaches to building the WBS?

83. If you are late, will anybody notice?

84. Is the duration of the program sufficient to ensure a cycle that will Autonomous car liability project the sustainability of the interventions?

85. What do they need to know about the Autonomous car liability project?

86. Did you read it correctly?

87. To what extent has the intervention strategy been adapted to the areas of intervention in which it is being implemented?

88. If task x starts two days late, what is the effect on the Autonomous car liability project end date?

89. How will users learn how to use the deliverables?

90. What factors are contributing to progress or delay in the achievement of products and results?

2.1 Project Management Plan: Autonomous car liability

91. Is mitigation authorized or recommended?

92. Are alternatives safe, functional, constructible, economical, reasonable and sustainable?

93. Why Change?

94. Who manages integration?

95. Do the proposed changes from the Autonomous car liability project include any significant risks to safety?

96. How well are you able to manage your risk?

97. When is a Autonomous car liability project management plan created?

98. Are comparable cost estimates used for comparing, screening and selecting alternative plans, and has a reasonable cost estimate been developed for the recommended plan?

99. What are the deliverables?

100. What goes into your Autonomous car liability project Charter?

101. Development trends and opportunities. What if the positive direction and vision of your organization

causes expected trends to change?

102. What did not work so well?

103. What worked well?

104. Is the budget realistic?

105. How do you organize the costs in the Autonomous car liability project management plan?

106. What data/reports/tools/etc. do your PMs need?

107. Is the engineering content at a feasibility level-of-detail, and is it sufficiently complete, to provide an adequate basis for the baseline cost estimate?

108. What if, for example, the positive direction and vision of your organization causes expected trends to change resulting in greater need than expected?

109. Was the peer (technical) review of the cost estimates duly coordinated with the cost estimate center of expertise and addressed in the review documentation and certification?

2.2 Scope Management Plan: Autonomous car liability

110. Has the Autonomous car liability project manager been identified?

111. Is there a requirements change management processes in place?

112. Is it possible to track all classes of Autonomous car liability project work (e.g. scheduled, un-scheduled, defect repair, etc.)?

113. Have the key functions and capabilities been defined and assigned to each release or iteration?

114. Have adequate procedures been put in place for Autonomous car liability project communication and status reporting across Autonomous car liability project boundaries (for example interdependent software development among interfacing systems)?

115. What is the estimated cost of creating and implementing?

116. Are target dates established for each milestone deliverable?

117. Has a structured approach been used to break work effort into manageable components (WBS)?

118. Are adequate resources provided for the quality assurance function?

119. Have the procedures for identifying budget variances been followed?

120. Are internal Autonomous car liability project status meetings held at reasonable intervals?

121. What weaknesses do you have?

122. Describe how the deliverables will be verified against the Autonomous car liability project scope. To whom will the deliverables be first presented for inspection and verification?

123. Do you have the reasons why the changes to your organizational systems and capabilities are required?

124. Is the Autonomous car liability project sponsor clearly communicating the business case or rationale for why this Autonomous car liability project is needed?

125. Are Autonomous car liability project leaders committed to this Autonomous car liability project full time?

126. Do you secure formal approval of changes and requirements from stakeholders?

127. Are you meeting with stake holders and team members?

128. Time estimation – how much time will be needed?

129. How do you plan to control Scope Creep?

2.3 Requirements Management Plan: Autonomous car liability

130. Is there formal agreement on who has authority to approve a change in requirements?

131. In case of software development; Should you have a test for each code module?

132. Will the contractors involved take full responsibility?

133. Are actual resources expenditures versus planned expenditures acceptable?

134. Will you document changes to requirements?

135. Is the system software (non-operating system) new to the IT Autonomous car liability project team?

136. Will you have access to stakeholders when you need them?

137. How will you communicate scheduled tasks to other team members?

138. If it exists, where is it housed?

139. How will the requirements become prioritized?

140. Who is responsible for quantifying the Autonomous car liability project requirements?

141. Will you use an assessment of the Autonomous car liability project environment as a tool to discover risk to the requirements process?

142. Define the help desk model. who will take full responsibility?

143. Who will do the reporting and to whom will reports be delivered?

144. Do you really need to write this document at all?

145. Did you distinguish the scope of work the contractor(s) will be required to do?

146. Is it new or replacing an existing business system or process?

147. What are you counting on?

148. What information regarding the Autonomous car liability project requirements will be reported?

149. How will bidders price evaluations be done, by deliverables, phases, or in a big bang?

2.4 Requirements Documentation: Autonomous car liability

150. Is your business case still valid?

151. What marketing channels do you want to use: e-mail, letter or sms?

152. Who is interacting with the system?

153. Basic work/business process; high-level, what is being touched?

154. Where do you define what is a customer, what are the attributes of customer?

155. Completeness. are all functions required by the customer included?

156. What is your Elevator Speech?

157. What is the risk associated with cost and schedule?

158. What will be the integration problems?

159. Are there legal issues?

160. Do your constraints stand?

161. Does the system provide the functions which best support the customers needs?

162. What is a show stopper in the requirements?

163. How does the proposed Autonomous car liability project contribute to the overall objectives of your organization?

164. What can tools do for us?

165. The problem with gathering requirements is right there in the word gathering. What images does it conjure?

166. What are current process problems?

167. What is effective documentation?

168. How will they be documented / shared?

169. Who is involved?

2.5 Requirements Traceability Matrix: Autonomous car liability

170. How do you manage scope?

171. Do you have a clear understanding of all subcontracts in place?

172. Will you use a Requirements Traceability Matrix?

173. Why do you manage scope?

174. Why use a WBS?

175. Describe the process for approving requirements so they can be added to the traceability matrix and Autonomous car liability project work can be performed. Will the Autonomous car liability project requirements become approved in writing?

176. What is the WBS?

177. What percentage of Autonomous car liability projects are producing traceability matrices between requirements and other work products?

178. Is there a requirements traceability process in place?

179. How will it affect the stakeholders personally in career?

180. What are the chronologies, contingencies,

consequences, criteria?

181. How small is small enough?

2.6 Project Scope Statement: Autonomous car liability

182. If you were to write a list of what should not be included in the scope statement, what are the things that you would recommend be described as out-of-scope?

183. If the scope changes, what will the impact be to your Autonomous car liability project in terms of duration, cost, quality, or any other important areas of the Autonomous car liability project?

184. Will the qa related information be reported regularly as part of the status reporting mechanisms?

185. Have the reports to be produced, distributed, and filed been defined?

186. Will this process be communicated to the customer and Autonomous car liability project team?

187. Will all Autonomous car liability project issues be unconditionally tracked through the issue resolution process?

188. Has everyone approved the Autonomous car liability projects scope statement?

189. How often do you estimate that the scope might change, and why?

190. What went right?

191. Is the Autonomous car liability project sponsor function identified and defined?

192. Is the quality function identified and assigned?

193. Is the plan under configuration management?

194. What went wrong?

195. Relevant - ask yourself can you get there; why are you doing this Autonomous car liability project?

196. Is there an information system for the Autonomous car liability project?

197. Are there issues that could affect the existing requirements for the result, service, or product if the scope changes?

198. Write a brief purpose statement for this Autonomous car liability project. Include a business justification statement. What is the product of this Autonomous car liability project?

199. Was planning completed before the Autonomous car liability project was initiated?

200. Autonomous car liability project lead, team lead, solution architect?

201. Does the scope statement still need some clarity?

2.7 Assumption and Constraint Log: Autonomous car liability

202. What to do at recovery?

203. Are formal code reviews conducted?

204. Is the definition of the Autonomous car liability project scope clear; what needs to be accomplished?

205. What would you gain if you spent time working to improve this process?

206. Is this process still needed?

207. Are funding and staffing resource estimates sufficiently detailed and documented for use in planning and tracking the Autonomous car liability project?

208. Can the requirements be traced to the appropriate components of the solution, as well as test scripts?

209. Have all stakeholders been identified?

210. Was the document/deliverable developed per the appropriate or required standards (for example, Institute of Electrical and Electronics Engineers standards)?

211. Is the steering committee active in Autonomous car liability project oversight?

212. What is positive about the current process?

213. Contradictory information between document sections?

214. After observing execution of process, is it in compliance with the documented Plan?

215. Does the traceability documentation describe the tool and/or mechanism to be used to capture traceability throughout the life cycle?

216. Do the requirements meet the standards of correctness, completeness, consistency, accuracy, and readability?

217. How are new requirements or changes to requirements identified?

218. Are there unnecessary steps that are creating bottlenecks and/or causing people to wait?

219. Model-building: what data-analytic strategies are useful when building proportional-hazards models?

2.8 Work Breakdown Structure: Autonomous car liability

220. How many levels?

221. How big is a work-package?

222. Why would you develop a Work Breakdown Structure?

223. How far down?

224. Who has to do it?

225. Do you need another level?

226. Is it still viable?

227. Can you make it?

228. When would you develop a Work Breakdown Structure?

229. When do you stop?

230. How will you and your Autonomous car liability project team define the Autonomous car liability projects scope and work breakdown structure?

231. Is the work breakdown structure (wbs) defined and is the scope of the Autonomous car liability project clear with assigned deliverable owners?

232. What has to be done?

233. When does it have to be done?

234. Is it a change in scope?

235. What is the probability that the Autonomous car liability project duration will exceed xx weeks?

236. Where does it take place?

2.9 WBS Dictionary: Autonomous car liability

237. Do the lines of authority for incurring indirect costs correspond to the lines of responsibility for management control of the same components of costs?

238. Are time-phased budgets established for planning and control of level of effort activity by category of resource; for example, type of manpower and/or material?

239. Are all authorized tasks assigned to identified organizational elements?

240. Are the variances between budgeted and actual indirect costs identified and analyzed at the level of assigned responsibility for control (indirect pool, department, etc.)?

241. Are the contractors estimates of costs at completion reconcilable with cost data reported to us?

242. Are all elements of indirect expense identified to overhead cost budgets of Autonomous car liability projections?

243. Are data elements reconcilable between internal summary reports and reports forwarded to us?

244. Changes in the direct base to which overhead

costs are allocated?

245. Is the anticipated (firm and potential) business base Autonomous car liability projected in a rational, consistent manner?

246. Are overhead budgets and costs being handled according to the disclosure statement when applicable, or otherwise properly classified (for example, engineering overhead, IR&D)?

247. Are overhead cost budgets established for each organization which has authority to incur overhead costs?

248. All cwbs elements specified for external reporting?

249. Are estimates of costs at completion generated in a rational, consistent manner?

250. Are retroactive changes to budgets for completed work specifically prohibited in an established procedure, and is this procedure adhered to?

251. Is authorization of budgets in excess of the contract budget base controlled formally and done with the full knowledge and recognition of the procuring activity?

252. Are the latest revised estimates of costs at completion compared with the established budgets at appropriate levels and causes of variances identified?

253. Is all budget available as management reserve identified and excluded from the performance measurement baseline?

254. Software specification, development, integration, and testing, licenses ?

255. Does the contractors system include procedures for measuring the performance of critical subcontractors?

2.10 Schedule Management Plan: Autonomous car liability

256. Is your organization certified as a supplier, wholesaler and/or regular dealer?

257. Is the development plan and/or process documented?

258. Goal: is the schedule feasible and at what cost?

259. Is there a procedure for management, control and release of schedule margin?

260. Does the ims include all contract and/or designated management control milestones?

261. List all schedule constraints here. Must the Autonomous car liability project be complete by a specified date?

262. How do you manage time?

263. What strengths do you have?

264. Are the results of quality assurance reviews provided to affected groups & individuals?

265. Are the key elements of a Autonomous car liability project Charter present?

266. Is a pmo (Autonomous car liability project management office) in place and provide oversight to

the Autonomous car liability project?

267. Have all involved Autonomous car liability project stakeholders and work groups committed to the Autonomous car liability project?

268. Is there an issues management plan in place?

269. Are all activities logically sequenced?

270. Are issues raised, assessed, actioned, and resolved in a timely and efficient manner?

271. Are Autonomous car liability project team members involved in detailed estimating and scheduling?

272. Have external dependencies been captured in the schedule?

273. Are the appropriate IT resources adequate to meet planned commitments?

274. Identify the amount of schedule variation that triggers a warning. What happens if a warning is triggered?

275. Has a quality assurance plan been developed for the Autonomous car liability project?

2.11 Activity List: Autonomous car liability

276. Should you include sub-activities?

277. When will the work be performed?

278. What went well?

279. How much slack is available in the Autonomous car liability project?

280. How can the Autonomous car liability project be displayed graphically to better visualize the activities?

281. The wbs is developed as part of a joint planning session. and how do you know that youhave done this right?

282. What will be performed?

283. Is there anything planned that does not need to be here?

284. What is the LF and LS for each activity?

285. How will it be performed?

286. What did not go as well?

287. How should ongoing costs be monitored to try to keep the Autonomous car liability project within budget?

288. How difficult will it be to do specific activities on this Autonomous car liability project?

289. Where will it be performed?

290. What are the critical bottleneck activities?

291. What is your organizations history in doing similar activities?

292. In what sequence?

2.12 Activity Attributes: Autonomous car liability

293. What activity do you think you should spend the most time on?

294. Activity: what is Missing?

295. What conclusions/generalizations can you draw from this?

296. What is missing?

297. Which method produces the more accurate cost assignment?

298. Activity: what is In the Bag?

299. Are the required resources available or need to be acquired?

300. How many resources do you need to complete the work scope within a limit of X number of days?

301. Is there a trend during the year?

302. Resources to accomplish the work?

303. Do you feel very comfortable with your prediction?

304. Would you consider either of corresponding activities an outlier?

305. Where else does it apply?

306. Have you identified the Activity Leveling Priority code value on each activity?

307. Resource is assigned to?

308. Are the required resources available?

309. Does your organization of the data change its meaning?

310. How difficult will it be to complete specific activities on this Autonomous car liability project?

2.13 Milestone List: Autonomous car liability

311. What are your competitors vulnerabilities?

312. Which path is the critical path?

313. How soon can the activity finish?

314. Insurmountable weaknesses?

315. What background experience, skills, and strengths does the team bring to your organization?

316. How late can each activity be finished and started?

317. Milestone pages should display the UserID of the person who added the milestone. Does a report or query exist that provides this audit information?

318. Identify critical paths (one or more) and which activities are on the critical path?

319. Political effects?

320. How will you get the word out to customers?

321. Do you foresee any technical risks or developmental challenges?

322. Calculate how long can activity be delayed?

323. Obstacles faced?

324. Sustaining internal capabilities?

325. Global influences?

326. Describe the concept of the technology, product or service that will be or has been developed. How will it be used?

327. How will the milestone be verified?

328. What specific improvements did you make to the Autonomous car liability project proposal since the previous time?

329. Usps (unique selling points)?

2.14 Network Diagram: Autonomous car liability

330. What job or jobs follow it?

331. What job or jobs could run concurrently?

332. Which type of network diagram allows you to depict four types of dependencies?

333. Where do schedules come from?

334. Are the gantt chart and/or network diagram updated periodically and used to assess the overall Autonomous car liability project timetable?

335. What are the Major Administrative Issues?

336. Will crashing x weeks return more in benefits than it costs?

337. Why must you schedule milestones, such as reviews, throughout the Autonomous car liability project?

338. If x is long, what would be the completion time if you break x into two parallel parts of y weeks and z weeks?

339. What to do and When?

340. Are you on time?

341. What activities must occur simultaneously with this activity?

342. What are the Key Success Factors?

343. What job or jobs precede it?

344. What activities must follow this activity?

345. How difficult will it be to do specific activities on this Autonomous car liability project?

346. What activity must be completed immediately before this activity can start?

347. Exercise: what is the probability that the Autonomous car liability project duration will exceed xx weeks?

2.15 Activity Resource Requirements: Autonomous car liability

348. Which logical relationship does the PDM use most often?

349. How many signatures do you require on a check and does this match what is in your policy and procedures?

350. How do you handle petty cash?

351. Other support in specific areas?

352. Organizational Applicability?

353. What is the Work Plan Standard?

354. Anything else?

355. Why do you do that?

356. When does monitoring begin?

357. Are there unresolved issues that need to be addressed?

358. Do you use tools like decomposition and rolling-wave planning to produce the activity list and other outputs?

359. What are constraints that you might find during the Human Resource Planning process?

360. Time for overtime?

2.16 Resource Breakdown Structure: Autonomous car liability

361. How should the information be delivered?

362. Any changes from stakeholders?

363. When do they need the information?

364. Why do you do it?

365. What is the primary purpose of the human resource plan?

366. What is the number one predictor of a groups productivity?

367. What is Autonomous car liability project communication management?

368. Is predictive resource analysis being done?

369. Who is allowed to see what data about which resources?

370. How can this help you with team building?

371. Who will use the system?

372. What is each stakeholders desired outcome for the Autonomous car liability project?

373. Who will be used as a Autonomous car liability

project team member?

374. What is the difference between % Complete and % work?

375. Why time management?

376. Which resources should be in the resource pool?

2.17 Activity Duration Estimates: Autonomous car liability

377. Did anything besides luck make a difference between success and failure?

378. Given your research into similar classes and the work you think is required for this Autonomous car liability project, what assumptions, variables, or costs would you change from the information provided above?

379. Do they make sense?

380. Is a work breakdown structure created to organize and to confirm the scope of each Autonomous car liability project?

381. Are reward and recognition systems defined to promote or reinforce desired behavior?

382. Autonomous car liability project manager has received activity duration estimates from his team. Which does one need in order to complete schedule development?

383. What are key inputs and outputs of the software?

384. Will it help promote wellness at your organization and reduce insurance costs?

385. List five reasons why organizations outsource. Why is there a growing trend in outsourcing,

especially in the government?

386. Are Autonomous car liability project activities decomposed into manageable components to ensure expected management control?

387. Are resource rates available to calculate Autonomous car liability project costs?

388. Why is there a new or renewed interest in the field of Autonomous car liability project management?

389. Do scope statements include the Autonomous car liability project objectives and expected deliverables?

390. After changes are approved are Autonomous car liability project documents updated and distributed?

391. Are operational definitions created to identify quality measurement criteria for specific activities?

392. After how many days will the lease cost be the same as the purchase cost for the equipment?

393. What is pmp certification, and why do you think the number of people earning it has grown so much in the past ten years?

394. What are the main parts of a scope statement?

395. Are team building activities completed to improve team performance?

2.18 Duration Estimating Worksheet: Autonomous car liability

396. For other activities, how much delay can be tolerated?

397. Do any colleagues have experience with your organization and/or RFPs?

398. Small or large Autonomous car liability project?

399. Define the work as completely as possible. What work will be included in the Autonomous car liability project?

400. Is this operation cost effective?

401. What utility impacts are there?

402. Done before proceeding with this activity or what can be done concurrently?

403. What is your role?

404. What is next?

405. When, then?

406. What work will be included in the Autonomous car liability project?

407. How can the Autonomous car liability project be displayed graphically to better visualize the activities?

408. What questions do you have?

409. What is an Average Autonomous car liability project?

410. Why estimate time and cost?

411. What info is needed?

412. Value pocket identification & quantification what are value pockets?

413. Is the Autonomous car liability project responsive to community need?

414. What is the total time required to complete the Autonomous car liability project if no delays occur?

2.19 Project Schedule: Autonomous car liability

415. How does a Autonomous car liability project get to be a year late ?

416. Activity charts and bar charts are graphical representations of a Autonomous car liability project schedule ...how do they differ?

417. How can slack be negative?

418. Have all Autonomous car liability project delays been adequately accounted for, communicated to all stakeholders and adjustments made in overall Autonomous car liability project schedule?

419. How much slack is available in the Autonomous car liability project?

420. Are all remaining durations correct?

421. If there are any qualifying green components to this Autonomous car liability project, what portion of the total Autonomous car liability project cost is green?

422. What does that mean?

423. Are procedures defined by which the Autonomous car liability project schedule may be changed?

424. How can you fix it?

425. Are activities connected because logic dictates the order in which others occur?

426. Is the structure for tracking the Autonomous car liability project schedule well defined and assigned to a specific individual?

427. Change management required?

428. Are the original Autonomous car liability project schedule and budget realistic?

429. How much detail?

430. Meet requirements?

431. If you can not fix it, how do you do it differently?

432. Why do you need to manage Autonomous car liability project Risk?

433. Why is software Autonomous car liability project disaster so common?

434. Autonomous car liability project work estimates Who is managing the work estimate quality of work tasks in the Autonomous car liability project schedule?

2.20 Cost Management Plan: Autonomous car liability

435. Are the payment terms being followed?

436. Are Autonomous car liability project leaders committed to this Autonomous car liability project full time?

437. Are written status reports provided on a designated frequent basis?

438. Designated small business reserve?

439. Is there a formal set of procedures supporting Issues Management?

440. Is it possible to track all classes of Autonomous car liability project work (e.g. scheduled, un-scheduled, defect repair, etc.)?

441. Has a Autonomous car liability project Communications Plan been developed?

442. What would you do differently what did not work?

443. Is there an approved case?

444. Quality assurance overheads?

445. Is the schedule updated on a periodic basis?

446. Was the scope definition used in task sequencing?

447. Are tasks tracked by hours?

448. Exclusions – is there scope to be performed or provided by others?

449. Is documentation created for communication with the suppliers and Vendors?

450. Are Autonomous car liability project contact logs kept up to date?

451. Are quality inspections and review activities listed in the Autonomous car liability project schedule(s)?

452. Is the Autonomous car liability project schedule available for all Autonomous car liability project team members to review?

453. Are enough systems & user personnel assigned to the Autonomous car liability project?

2.21 Activity Cost Estimates: Autonomous car liability

454. Did the Autonomous car liability project team have the right skills?

455. What do you want to know about the stay to know if costs were inappropriately high or low?

456. How do you do activity recasts?

457. Does the estimator have experience?

458. How difficult will it be to do specific tasks on the Autonomous car liability project?

459. In which phase of the acquisition process cycle does source qualifications reside?

460. Does the activity rely on a common set of tools to carry it out?

461. Where can you get activity reports?

462. Padding is bad and contingencies are good. what is the difference?

463. Why do you manage cost?

464. Was the consultant knowledgeable about the program?

465. What is the estimators estimating history?

466. Can you delete activities or make them inactive?

467. Specific - is the objective clear in terms of what, how, when, and where the situation will be changed?

468. What makes a good activity description?

469. What makes a good expected result statement?

470. What were things that you did well, and could improve, and how?

471. Performance bond should always provide what part of the contract value?

472. How do you change activities?

2.22 Cost Estimating Worksheet: Autonomous car liability

473. Identify the timeframe necessary to monitor progress and collect data to determine how the selected measure has changed?

474. What will others want?

475. What costs are to be estimated?

476. Ask: are others positioned to know, are others credible, and will others cooperate?

477. Is the Autonomous car liability project responsive to community need?

478. How will the results be shared and to whom?

479. Will the Autonomous car liability project collaborate with the local community and leverage resources?

480. What can be included?

481. Is it feasible to establish a control group arrangement?

482. What happens to any remaining funds not used?

483. Can a trend be established from historical performance data on the selected measure and are the criteria for using trend analysis or forecasting

methods met?

484. What is the estimated labor cost today based upon this information?

485. What is the purpose of estimating?

486. Does the Autonomous car liability project provide innovative ways for stakeholders to overcome obstacles or deliver better outcomes?

487. Who is best positioned to know and assist in identifying corresponding factors?

488. What additional Autonomous car liability project(s) could be initiated as a result of this Autonomous car liability project?

2.23 Cost Baseline: Autonomous car liability

489. On budget?

490. What does a good WBS NOT look like?

491. What is your organizations history in doing similar tasks?

492. Does a process exist for establishing a cost baseline to measure Autonomous car liability project performance?

493. What is cost and Autonomous car liability project cost management?

494. Are you asking management for something as a result of this update?

495. Definition of done can be traced back to the definitions of what are you providing to the customer in terms of deliverables?

496. Have the resources used by the Autonomous car liability project been reassigned to other units or Autonomous car liability projects?

497. Escalation criteria met?

498. Have you identified skills that are missing from your team?

499. How will cost estimates be used?

500. Autonomous car liability project goals -should others be reconsidered?

501. Are there contingencies or conditions related to the acceptance?

502. Have all the product or service deliverables been accepted by the customer?

503. What do you want to measure ?

504. Where do changes come from?

505. Will the Autonomous car liability project fail if the change request is not executed?

506. Is request in line with priorities?

507. Has the Autonomous car liability project (or Autonomous car liability project phase) been evaluated against each objective established in the product description and Integrated Autonomous car liability project Plan?

508. Is there anything unique in this Autonomous car liability projects scope statement that will affect resources?

2.24 Quality Management Plan: Autonomous car liability

509. What process do you use to minimize errors, defects, and rework?

510. Have all necessary approvals been obtained?

511. How are your organizations compensation and recognition approaches and the performance management system used to reinforce high performance?

512. Are there ways to reduce the time it takes to get something approved?

513. Does the plan conform to standards?

514. Modifications to the requirements?

515. Does the program use other agents to collect samples?

516. Does a documented Autonomous car liability project organizational policy & plan (i.e. governance model) exist?

517. What are the appropriate test methods to be used?

518. Have adequate resources been provided by management to ensure Autonomous car liability project success?

519. Are there trends or hot spots?

520. Contradictory information between different documents?

521. How do you ensure that protocols are up to date?

522. How are records kept in the office?

523. How does your organization recruit, hire, and retain new employees?

524. Why quality management?

525. How do you ensure that your sampling methods and procedures meet your data quality objectives?

526. How does your organization maintain a safe and healthy work environment?

527. Are you following the quality standards?

2.25 Quality Metrics: Autonomous car liability

528. What does this tell us?

529. When is the security analysis testing complete?

530. Who is willing to lead?

531. What are your organizations next steps?

532. Do the operators focus on determining; is there anything you need to worry about?

533. What forces exist that would cause them to change?

534. Did the team meet the Autonomous car liability project success criteria documented in the Quality Metrics Matrix?

535. How exactly do you define when differences exist?

536. Has trace of defects been initiated?

537. What is the timeline to meet your goal?

538. What method of measurement do you use?

539. What level of statistical confidence do you use?

540. Are documents on hand to provide explanations

of privacy and confidentiality?

541. Is there a set of procedures to capture, analyze and act on quality metrics?

542. Has it met internal or external standards?

543. Have alternatives been defined in the event that failure occurs?

544. How do you measure?

545. Are applicable standards referenced and available?

546. Do you know how much profit a 10% decrease in waste would generate?

547. Has risk analysis been adequately reviewed?

2.26 Process Improvement Plan: Autonomous car liability

548. Has a process guide to collect the data been developed?

549. Does your process ensure quality?

550. Has the time line required to move measurement results from the points of collection to databases or users been established?

551. What makes people good SPI coaches?

552. If a process improvement framework is being used, which elements will help the problems and goals listed?

553. The motive is determined by asking, Why do you want to achieve this goal?

554. Are you making progress on the goals?

555. What is the test-cycle concept?

556. Where do you focus?

557. Where do you want to be?

558. What lessons have you learned so far?

559. How do you manage quality?

560. To elicit goal statements, do you ask a question such as, What do you want to achieve?

561. What personnel are the sponsors for that initiative?

562. Modeling current processes is great, and will you ever see a return on that investment?

563. What personnel are the coaches for your initiative?

564. Why do you want to achieve the goal?

2.27 Responsibility Assignment Matrix: Autonomous car liability

565. Identify and isolate causes of favorable and unfavorable cost and schedule variances?

566. What tool can show you individual and group allocations?

567. Are there any drawbacks to using a responsibility assignment matrix?

568. The already stated responsible for overhead performance control of related costs?

569. The total budget for the contract (including estimates for authorized and unpriced work)?

570. Are management actions taken to reduce indirect costs when there are significant adverse variances?

571. What expertise is not available in your department?

572. Is the entire contract planned in time-phased control accounts to the extent practicable?

573. Are people encouraged to bring up issues?

574. Are your organizations and items of cost assigned to each pool identified?

575. Are indirect costs charged to the appropriate indirect pools and incurring organization?

576. Most people let you know when others re too busy, and are others really too busy?

577. Will too many Communicating responsibilities tangle the Autonomous car liability project in unnecessary communications?

578. Are meaningful indicators identified for use in measuring the status of cost and schedule performance?

579. Evaluate the impact of schedule changes, work around, etc?

580. Evaluate the performance of operating organizations?

581. Past experience – the person or the group worked at something similar in the past?

582. The already stated responsible for the establishment of budgets and assignment of resources for overhead performance?

583. Is work progressively subdivided into detailed work packages as requirements are defined?

2.28 Roles and Responsibilities: Autonomous car liability

584. What should you do now to prepare for your career 5+ years from now?

585. Implementation of actions: Who are the responsible units?

586. What should you do now to ensure that you are exceeding expectations and excelling in your current position?

587. Attainable / achievable: the goal is attainable; can you actually accomplish the goal?

588. Key conclusions and recommendations: Are conclusions and recommendations relevant and acceptable?

589. Is the data complete?

590. What specific behaviors did you observe?

591. Are your policies supportive of a culture of quality data?

592. What areas would you highlight for changes or improvements?

593. What should you do now to prepare yourself for a promotion, increased responsibilities or a different job?

594. Concern: where are you limited or have no authority, where you can not influence?

595. Once the responsibilities are defined for the Autonomous car liability project, have the deliverables, roles and responsibilities been clearly communicated to every participant?

596. Does your vision/mission support a culture of quality data?

597. Are your budgets supportive of a culture of quality data?

598. Are Autonomous car liability project team roles and responsibilities identified and documented?

599. What should you do now to ensure that you are meeting all expectations of your current position?

600. Authority: what areas/Autonomous car liability projects in your work do you have the authority to decide upon and act on the already stated decisions?

601. Does the team have access to and ability to use data analysis tools?

602. Have you ever been a part of this team?

2.29 Human Resource Management Plan: Autonomous car liability

603. Are mitigation strategies identified?

604. Are Autonomous car liability project team roles and responsibilities identified and documented?

605. Is there an on-going process in place to monitor Autonomous car liability project risks?

606. Does the Autonomous car liability project have a formal Autonomous car liability project Charter?

607. How are superior performers differentiated from average performers?

608. Is it standard practice to formally commit stakeholders to the Autonomous car liability project via agreements?

609. Are change requests logged and managed?

610. Responsiveness to change and the resulting demands for different skills and abilities?

611. Is the assigned Autonomous car liability project manager a PMP (Certified Autonomous car liability project manager) and experienced?

612. Are milestone deliverables effectively tracked and compared to Autonomous car liability project plan?

613. List the assumptions made to date. What did you have to assume to be true to complete the charter?

614. Have Autonomous car liability project management standards and procedures been identified / established and documented?

615. How will the Autonomous car liability project manage expectations & meet needs and requirements?

616. Has the schedule been baselined?

617. Have lessons learned been conducted after each Autonomous car liability project release?

618. Were Autonomous car liability project team members involved in detailed estimating and scheduling?

619. Quality of people required to meet the forecast needs of the department?

620. Have adequate resources been provided by management to ensure Autonomous car liability project success?

621. Is there general agreement & acceptance of the current status and progress of the Autonomous car liability project?

2.30 Communications Management Plan: Autonomous car liability

622. Who to learn from?

623. Which team member will work with each stakeholder?

624. Is there an important stakeholder who is actively opposed and will not receive messages?

625. Can you think of other people who might have concerns or interests?

626. What is the stakeholders level of authority?

627. Are there potential barriers between the team and the stakeholder?

628. How is this initiative related to other portfolios, programs, or Autonomous car liability projects?

629. What are the interrelationships?

630. What approaches do you use?

631. Why do you manage communications?

632. Are others needed?

633. Who will use or be affected by the result of a Autonomous car liability project?

634. What is Autonomous car liability project communications management?

635. Which stakeholders are thought leaders, influences, or early adopters?

636. Who did you turn to if you had questions?

637. What data is going to be required?

638. Do you have members of your team responsible for certain stakeholders?

639. Who is responsible?

640. Who is the stakeholder?

2.31 Risk Management Plan: Autonomous car liability

641. Market risk -will the new service or product be useful to your organization or marketable to others?

642. Degree of confidence in estimated size estimate?

643. For software; does the software interface with new or unproven hardware or unproven vendor products?

644. How is risk response planning performed?

645. Have top software and customer managers formally committed to support the Autonomous car liability project?

646. User involvement: do you have the right users?

647. Are people attending meetings and doing work?

648. What are the cost, schedule and resource impacts of avoiding the risk?

649. Do you have a consistent repeatable process that is actually used?

650. What things might go wrong?

651. Are you on schedule?

652. How can the process be made more effective or

less cumbersome (process improvements)?

653. What is the likelihood?

654. Are end-users enthusiastically committed to the Autonomous car liability project and the system/ product to be built?

655. Do you train all developers in the process?

656. Risk documentation: what reporting formats and processes will be used for risk management activities?

657. Anticipated volatility of the requirements?

658. What would you do differently?

659. For software; are compilers and code generators available and suitable for the product to be built?

660. Is a software Autonomous car liability project management tool available?

2.32 Risk Register: Autonomous car liability

661. How well are risks controlled?

662. Who is accountable?

663. What will be done?

664. Are there any gaps in the evidence?

665. What should you do when?

666. What should you do now?

667. Are corrective measures implemented as planned?

668. Which key risks have ineffective responses or outstanding improvement actions?

669. Who needs to know about this?

670. Risk probability and impact: how will the probabilities and impacts of risk items be assessed?

671. What has changed since the last period?

672. What action, if any, has been taken to respond to the risk?

673. Cost/benefit – how much will the proposed mitigations cost and how does this cost compare with

the potential cost of the risk event/situation should it occur?

674. Manageability – have mitigations to the risk been identified?

675. What should the audit role be in establishing a risk management process?

676. Can the likelihood and impact of failing to achieve corresponding recommendations and action plans be assessed?

677. Methodology: how will risk management be performed on this Autonomous car liability project?

678. What may happen or not go according to plan?

679. When will it happen?

2.33 Probability and Impact Assessment: Autonomous car liability

680. Are testing tools available and suitable?

681. Monitoring of the overall Autonomous car liability project status - are there any changes in the Autonomous car liability project that can effect and cause new possible risks?

682. Who are the international/overseas Autonomous car liability project partners (equipment supplier/ supplier/consultant/contractor) for this Autonomous car liability project?

683. Is a software Autonomous car liability project management tool available?

684. How realistic is the timing of introduction?

685. Are tool mentors available?

686. What are the preparations required for facing difficulties?

687. Are requirements fully understood by the software engineering team and customers?

688. What is the past performance of the Autonomous car liability project manager?

689. Is the process supported by tools?

690. Does the software interface with new or unproven hardware or unproven vendor products?

691. Can you avoid altogether some things that might go wrong?

692. Which role do you have in the Autonomous car liability project?

693. Are team members trained in the use of the tools?

694. Do end-users have realistic expectations?

695. Can the Autonomous car liability project proceed without assuming the risk?

696. Why has this particular mode of contracting been chosen?

697. What should be the external organizations responsibility vis-à-vis total stake in the Autonomous car liability project?

698. How would you assess the risk management process in the Autonomous car liability project?

699. Are the facilities, expertise, resources, and management know-how available to handle the situation?

2.34 Probability and Impact Matrix: Autonomous car liability

700. During Autonomous car liability project executing, a team member identifies a risk that is not in the risk register. What should you do?

701. What are the methods to deal with risks?

702. Are Autonomous car liability project requirements stable?

703. Is the number of people on the Autonomous car liability project team adequate to do the job?

704. What is the probability of the risk occurring?

705. What can you use the analyzed risks for?

706. Are the risk data complete?

707. What changes in the regulation are forthcoming?

708. What are its business ethics?

709. What are the ways you measure and evaluate risks?

710. Which should be probably done NEXT?

711. Which phase of the Autonomous car liability project do you take part in?

712. Several experts are offsite, and wish to be included. How can this be done?

713. Are there new risks that mitigation strategies might introduce?

714. How do you analyze the risks in the different types of Autonomous car liability projects?

715. How well is the risk understood?

716. How to prioritize risks?

717. Do you have a mechanism for managing change?

2.35 Risk Data Sheet: Autonomous car liability

718. During work activities could hazards exist?

719. Are new hazards created?

720. What can happen?

721. How can hazards be reduced?

722. What was measured?

723. What are your core values?

724. What will be the consequences if it happens?

725. How do you handle product safely?

726. What if client refuses?

727. What is the environment within which you operate (social trends, economic, community values, broad based participation, national directions etc.)?

728. What do you know?

729. What are you trying to achieve (Objectives)?

730. How reliable is the data source?

731. Type of risk identified?

732. Whom do you serve (customers)?

733. What are you weak at and therefore need to do better?

734. Will revised controls lead to tolerable risk levels?

735. What are the main threats to your existence?

2.36 Procurement Management Plan: Autonomous car liability

736. Is it possible to track all classes of Autonomous car liability project work (e.g. scheduled, un-scheduled, defect repair, etc.)?

737. Are decisions captured in a decisions log?

738. Is stakeholder involvement adequate?

739. Similar Autonomous car liability projects?

740. Is the structure for tracking the Autonomous car liability project schedule well defined and assigned to a specific individual?

741. Are stakeholders aware and supportive of the principles and practices of modern software estimation?

742. What communication items need improvement?

743. Are internal Autonomous car liability project status meetings held at reasonable intervals?

744. Are actuals compared against estimates to analyze and correct variances?

745. Are Autonomous car liability project leaders committed to this Autonomous car liability project full time?

746. Has a quality assurance plan been developed for the Autonomous car liability project?

747. Autonomous car liability project Objectives?

748. Are governance roles and responsibilities documented?

749. Are quality metrics defined?

750. Has the budget been baselined?

751. Has a Autonomous car liability project Communications Plan been developed?

2.37 Source Selection Criteria: Autonomous car liability

752. How are clarifications and communications appropriately used?

753. What is the basis of an estimate and what assumptions were made?

754. What is the last item a Autonomous car liability project manager must do to finalize Autonomous car liability project close-out?

755. How should the oral presentations be handled?

756. In the technical/management area, what criteria do you use to determine the final evaluation ratings?

757. What is the role of counsel in the procurement process?

758. What aspects should the contracting officer brief the Autonomous car liability project on prior to evaluation of proposals?

759. What should a Draft Request for Proposal (DRFP) include?

760. What information is to be provided and when should it be provided?

761. What should a DRFP include?

762. How do you encourage efficiency and consistency?

763. Do you consider all weaknesses, significant weaknesses, and deficiencies?

764. What should communications be used to accomplish?

765. Is this a cost contract?

766. How should the preproposal conference be conducted?

767. Are there any common areas of weaknesses or deficiencies in the proposals in the competitive range?

768. Is the contracting office likely to receive more purchase requests for this item or service during the coming year?

769. When must you conduct a debriefing?

770. Do you prepare an independent cost estimate?

771. Who is entitled to a debriefing?

2.38 Stakeholder Management Plan: Autonomous car liability

772. Which impacts could serve as impediments?

773. What inspection and testing is to be performed?

774. Were Autonomous car liability project team members involved in detailed estimating and scheduling?

775. Are Autonomous car liability project contact logs kept up to date?

776. Are metrics used to evaluate and manage Vendors?

777. Are there standards for code development?

778. Have stakeholder accountabilities & responsibilities been clearly defined?

779. What are reporting requirements?

780. Who might be involved in developing a charter?

781. Who is responsible for accepting the reports produced by the process?

782. Are all resource assumptions documented?

783. Is there an onboarding process in place?

784. Is there a Steering Committee in place?

785. How will you engage this stakeholder and gain commitment?

786. Alignment to strategic goals & objectives?

787. Are trade-offs between accepting the risk and mitigating the risk identified?

2.39 Change Management Plan: Autonomous car liability

788. Who might present the most resistance?

789. Do you need a new organization structure?

790. What risks may occur upfront?

791. Has the training provider been established?

792. Is there a software application relevant to this deliverable?

793. Has the priority for this Autonomous car liability project been set by the Business Unit Management Team?

794. Identify the current level of skills and knowledge and behaviours of the group that will be impacted on. What prerequisite knowledge do corresponding groups need?

795. Has an information & communications plan been developed?

796. What are the responsibilities assigned to each role?

797. Will the culture embrace or reject this change?

798. How many people are required in each of the roles?

799. What are you trying to achieve as a result of communication?

800. What new roles are needed?

801. How prevalent is Resistance to Change?

802. Has the relevant business unit been notified of installation and support requirements?

803. Has the training co-ordinator been provided with the training details and put in place the necessary arrangements?

804. What are the specific target groups / audience that will be impacted by this change?

805. What are the major changes to processes?

806. How badly can information be misinterpreted?

3.0 Executing Process Group: Autonomous car liability

807. Do Autonomous car liability project managers understand your organizational context for Autonomous car liability projects?

808. How can you use Microsoft Autonomous car liability project and Excel to assist in Autonomous car liability project risk management?

809. Does the Autonomous car liability project team have the right skills?

810. How do you prevent staff are just doing busywork to pass the time?

811. Do schedule issues conflicts?

812. Is the Autonomous car liability project making progress in helping to achieve the set results?

813. Are the necessary foundations in place to ensure the sustainability of the results of the programme?

814. What is the critical path for this Autonomous car liability project and how long is it?

815. What does it mean to take a systems view of a Autonomous car liability project?

816. What areas does the group agree are the biggest success on the Autonomous car liability project?

817. When do you share the scorecard with managers?

818. What were things that you did very well and want to do the same again on the next Autonomous car liability project?

819. How many different communication channels does the Autonomous car liability project team have?

820. How well defined and documented were the Autonomous car liability project management processes you chose to use?

821. Are decisions made in a timely manner?

822. What type of people would you want on your team?

823. Will outside resources be needed to help?

3.1 Team Member Status Report: Autonomous car liability

824. Does the product, good, or service already exist within your organization?

825. Why is it to be done?

826. How it is to be done?

827. Are your organizations Autonomous car liability projects more successful over time?

828. Are the attitudes of staff regarding Autonomous car liability project work improving?

829. Do you have an Enterprise Autonomous car liability project Management Office (EPMO)?

830. Is there evidence that staff is taking a more professional approach toward management of your organizations Autonomous car liability projects?

831. Will the staff do training or is that done by a third party?

832. When a teams productivity and success depend on collaboration and the efficient flow of information, what generally fails them?

833. How can you make it practical?

834. The problem with Reward & Recognition

Programs is that the truly deserving people all too often get left out. How can you make it practical?

835. Does every department have to have a Autonomous car liability project Manager on staff?

836. How much risk is involved?

837. What specific interest groups do you have in place?

838. How will resource planning be done?

839. What is to be done?

840. Does your organization have the means (staff, money, contract, etc.) to produce or to acquire the product, good, or service?

841. Are the products of your organizations Autonomous car liability projects meeting customers objectives?

842. How does this product, good, or service meet the needs of the Autonomous car liability project and your organization as a whole?

3.2 Change Request: Autonomous car liability

843. Will the change use memory to the extent that other functions will be not have sufficient memory to operate effectively?

844. Has your address changed?

845. Why were your requested changes rejected or not made?

846. What can be filed?

847. Will all change requests and current status be logged?

848. How do you get changes (code) out in a timely manner?

849. What mechanism is used to appraise others of changes that are made?

850. Does the schedule include Autonomous car liability project management time and change request analysis time?

851. What is the purpose of change control?

852. Have scm procedures for noting the change, recording it, and reporting it been followed?

853. Who will perform the change?

854. Should staff call into the helpdesk or go to the website?

855. How are the measures for carrying out the change established?

856. Why control change across the life cycle?

857. Are there requirements attributes that can discriminate between high and low reliability?

858. What are the duties of the change control team?

859. How is the change documented (format, content, storage)?

860. Since there are no change requests in your Autonomous car liability project at this point, what must you have before you begin?

861. Who needs to approve change requests?

862. What needs to be communicated?

3.3 Change Log: Autonomous car liability

863. Is this a mandatory replacement?

864. How does this relate to the standards developed for specific business processes?

865. When was the request approved?

866. Do the described changes impact on the integrity or security of the system?

867. How does this change affect the timeline of the schedule?

868. Does the suggested change request represent a desired enhancement to the products functionality?

869. Is the change request open, closed or pending?

870. When was the request submitted?

871. Is the requested change request a result of changes in other Autonomous car liability project(s)?

872. Should a more thorough impact analysis be conducted?

873. Is the submitted change a new change or a modification of a previously approved change?

874. Is the change backward compatible without

limitations?

875. Who initiated the change request?

876. How does this change affect scope?

877. Does the suggested change request seem to represent a necessary enhancement to the product?

878. Is the change request within Autonomous car liability project scope?

879. Will the Autonomous car liability project fail if the change request is not executed?

3.4 Decision Log: Autonomous car liability

880. How do you define success?

881. What makes you different or better than others companies selling the same thing?

882. Decision-making process; how will the team make decisions?

883. What was the rationale for the decision?

884. Who is the decisionmaker?

885. Adversarial environment. is your opponent open to a non-traditional workflow, or will it likely challenge anything you do?

886. How effective is maintaining the log at facilitating organizational learning?

887. Who will be given a copy of this document and where will it be kept?

888. How consolidated and comprehensive a story can you tell by capturing currently available incident data in a central location and through a log of key decisions during an incident?

889. Which variables make a critical difference?

890. Is your opponent open to a non-traditional

workflow, or will it likely challenge anything you do?

891. At what point in time does loss become unacceptable?

892. Do strategies and tactics aimed at less than full control reduce the costs of management or simply shift the cost burden?

893. Does anything need to be adjusted?

894. How do you know when you are achieving it?

895. Meeting purpose; why does this team meet?

896. What eDiscovery problem or issue did your organization set out to fix or make better?

897. What is your overall strategy for quality control / quality assurance procedures?

898. How does the use a Decision Support System influence the strategies/tactics or costs?

899. What is the average size of your matters in an applicable measurement?

3.5 Quality Audit: Autonomous car liability

900. Does the report read coherently?

901. Do all staff have the necessary authority and resources to deliver what is expected of them?

902. Is there a written procedure for receiving materials?

903. How does your organization know that its system for managing intellectual property issues is appropriately effective, constructive and fair?

904. Is there a written corporate quality policy?

905. What are you trying to do?

906. Why are you trying to do it?

907. How does your organization know that its range of activities are being reviewed as rigorously and constructively as they could be?

908. Does everyone know what they are supposed to be doing, how and why?

909. Are the policies and processes, as set out in the Quality Audit Manual, properly applied?

910. What are your supplier audits?

911. How do you indicate the extent to which your personnel would be expected to contribute to the work effort?

912. How does your organization know that its security arrangements are appropriately effective and constructive?

913. How does your organization know that its system for governing staff behaviour is appropriately effective and constructive?

914. Is refuse and garbage adequately stored and disposed of with sufficient frequency to prevent contamination?

915. What is your organizations greatest strength?

916. How does your organization know that its methods are appropriately effective and constructive?

917. How does your organization know that the support for its staff is appropriately effective and constructive?

918. How does your organization know that its system for ensuring that its training activities are appropriately resourced and support is appropriately effective and constructive?

919. How do you know what, specifically, is required of you in your work?

3.6 Team Directory: Autonomous car liability

920. When will you produce deliverables?

921. Who will write the meeting minutes and distribute?

922. Process decisions: which organizational elements and which individuals will be assigned management functions?

923. Do purchase specifications and configurations match requirements?

924. Who should receive information (all stakeholders)?

925. Where should the information be distributed?

926. Who are your stakeholders (customers, sponsors, end users, team members)?

927. Process decisions: is work progressing on schedule and per contract requirements?

928. Timing: when do the effects of communication take place?

929. Decisions: is the most suitable form of contract being used?

930. Does a Autonomous car liability project

team directory list all resources assigned to the Autonomous car liability project?

931. Process decisions: do job conditions warrant additional actions to collect job information and document on-site activity?

932. Contract requirements complied with?

933. Days from the time the issue is identified?

934. Where will the product be used and/or delivered or built when appropriate?

935. Decisions: what could be done better to improve the quality of the constructed product?

936. Who will talk to the customer?

937. How does the team resolve conflicts and ensure tasks are completed?

3.7 Team Operating Agreement: Autonomous car liability

938. Do you solicit member feedback about meetings and what would make them better?

939. Must your members collaborate successfully to complete Autonomous car liability projects?

940. Reimbursements: how will the team members be reimbursed for expenses and time commitments?

941. What individual strengths does each team member bring to the group?

942. Are there more than two native languages represented by your team?

943. Are there more than two national cultures represented by your team?

944. Did you determine the technology methods that best match the messages to be communicated?

945. Do you prevent individuals from dominating the meeting?

946. Are there influences outside the team that may affect performance, and if so, have you identified and addressed them?

947. What are some potential sources of conflict among team members?

948. Confidentiality: how will confidential information be handled?

949. What types of accommodations will be formulated and put in place for sustaining the team?

950. The method to be used in the decision making process; Will it be consensus, majority rule, or the supervisor having the final say?

951. Do you use a parking lot for any items that are important and outside of the agenda?

952. Has the appropriate access to relevant data and analysis capability been granted?

953. What resources can be provided for the team in terms of equipment, space, time for training, protected time and space for meetings, and travel allowances?

954. Did you draft the meeting agenda?

955. How will group handle unplanned absences?

956. Did you recap the meeting purpose, time, and expectations?

957. To whom do you deliver your services?

3.8 Team Performance Assessment: Autonomous car liability

958. To what degree does the teams approach to its work allow for modification and improvement over time?

959. To what degree will team members, individually and collectively, commit time to help themselves and others learn and develop skills?

960. To what degree can team members frequently and easily communicate with one another?

961. If you have criticized someones work for method variance in your role as reviewer, what was the circumstance?

962. To what degree are sub-teams possible or necessary?

963. To what degree does the teams work approach provide opportunity for members to engage in fact-based problem solving?

964. To what degree do members articulate the goals beyond the team membership?

965. To what degree can team members vigorously define the teams purpose in considerations with others who are not part of the functioning team?

966. What is method variance?

967. Do friends perform better than acquaintances?

968. Does more radicalness mean more perceived benefits?

969. How much interpersonal friction is there in your team?

970. To what degree are staff involved as partners in the improvement process?

971. To what degree do members understand and articulate the same purpose without relying on ambiguous abstractions?

972. Lack of method variance in self-reported affect and perceptions at work: Reality or artifact?

973. To what degree are the teams goals and objectives clear, simple, and measurable?

974. Can familiarity breed backup?

975. Do you give group members authority to make at least some important decisions?

976. To what degree is there a sense that only the team can succeed?

977. To what degree will the team ensure that all members equitably share the work essential to the success of the team?

3.9 Team Member Performance Assessment: Autonomous car liability

978. Which training platform formats (i.e., mobile, virtual, videogame-based) were implemented in your effort(s)?

979. Is it critical or vital to the job?

980. Why do performance reviews?

981. What variables that affect team members achievement are within your control?

982. What stakeholders must be involved in the development and oversight of the performance plan?

983. What is the large, desired outcome?

984. What entity leads the process, selects a potential restructuring option and develops the plan?

985. What are best practices for delivering and developing training evaluations to maximize the benefits of leveraging emerging technologies?

986. What is a significant fact or event?

987. To what degree are the relative importance and priority of the goals clear to all team members?

988. What changes do you need to make to align practices with beliefs?

989. How are training activities developed from a technical perspective?

990. To what degree do team members understand one anothers roles and skills?

991. Are any governance changes sufficient to impact achievement?

992. What are the key duties or tasks of the Ratee?

993. How accurately is your plan implemented?

994. Where can team members go for more detailed information on performance measurement and assessment?

3.10 Issue Log: Autonomous car liability

995. How is this initiative related to other portfolios, programs, or Autonomous car liability projects?

996. Are you constantly rushing from meeting to meeting?

997. Who reported the issue?

998. What is a Stakeholder?

999. Do you feel a register helps?

1000. What effort will a change need?

1001. Who do you turn to if you have questions?

1002. Persistence; will users learn a work around or will they be bothered every time?

1003. Do you often overlook a key stakeholder or stakeholder group?

1004. What is the status of the issue?

1005. Is access to the Issue Log controlled?

1006. How often do you engage with stakeholders?

1007. In your work, how much time is spent on stakeholder identification?

1008. What steps can you take for positive relationships?

1009. Why multiple evaluators?

1010. Why not more evaluators?

1011. Who were proponents/opponents?

4.0 Monitoring and Controlling Process Group: Autonomous car liability

1012. What is the timeline for the Autonomous car liability project?

1013. What do they need to know about the Autonomous car liability project?

1014. Where is the Risk in the Autonomous car liability project?

1015. Mitigate. what will you do to minimize the impact should a risk event occur?

1016. Contingency planning. if a risk event occurs, what will you do?

1017. In what way has the program come up with innovative measures for problem-solving?

1018. If a risk event occurs, what will you do?

1019. Does the solution fit in with organizations technical architectural requirements?

1020. Who needs to be involved in the planning?

1021. Change, where should you look for problems?

1022. How is agile portfolio management done?

1023. Based on your Autonomous car liability project communication management plan, what worked well?

1024. Is there undesirable impact on staff or resources?

1025. Are the services being delivered?

1026. Where is the Risk in the Autonomous car liability project?

4.1 Project Performance Report: Autonomous car liability

1027. To what degree do team members articulate the teams work approach?

1028. To what degree are the demands of the task compatible with and converge with the relationships of the informal organization?

1029. To what degree can the team measure progress against specific goals?

1030. To what degree does the team possess adequate membership to achieve its ends?

1031. To what degree do team members frequently explore the teams purpose and its implications?

1032. To what degree can team members meet frequently enough to accomplish the teams ends?

1033. How is the data used?

1034. How will procurement be coordinated with other Autonomous car liability project aspects, such as scheduling and performance reporting?

1035. To what degree does the funding match the requirement?

1036. To what degree will the approach capitalize on and enhance the skills of all team members in a manner that takes into consideration other demands on members of the team?

1037. What is the degree to which rules govern information exchange between groups?

1038. To what degree does the teams work approach provide opportunity for members to engage in results-based evaluation?

1039. Next Steps?

1040. To what degree do individual skills and abilities match task demands?

1041. To what degree do the relationships of the informal organization motivate taskrelevant behavior and facilitate task completion?

1042. To what degree will the team adopt a concrete, clearly understood, and agreed-upon approach that will result in achievement of the teams goals?

4.2 Variance Analysis: Autonomous car liability

1043. How do you verify authorization to proceed with all authorized work?

1044. How do you identify and isolate causes of favorable and unfavorable cost and schedule variances?

1045. How does the monthly budget compare to the actual experience?

1046. Historical experience?

1047. Does the accounting system provide a basis for auditing records of direct costs chargeable to the contract?

1048. Are there externalities from having some customers, even if they are unprofitable in the short run?

1049. What is the budgeted cost for work scheduled?

1050. What is the actual cost of work performed?

1051. Are there changes in the overhead pool and/or organization structures?

1052. Why do variances exist?

1053. Does the scheduling system identify in a timely

manner the status of work?

1054. Are all budgets assigned to control accounts?

1055. There are detailed schedules which support control account and work package start and completion dates/events?

1056. Are data elements reconcilable between internal summary reports and reports forwarded to the stakeholders?

1057. Who are responsible for overhead performance control of related costs?

1058. Do work packages consist of discrete tasks which are adequately described?

1059. What is the total budget for the Autonomous car liability project (including estimates for authorized and unpriced work)?

1060. Who are responsible for the establishment of budgets and assignment of resources for overhead performance?

4.3 Earned Value Status: Autonomous car liability

1061. Are you hitting your Autonomous car liability projects targets?

1062. If earned value management (EVM) is so good in determining the true status of a Autonomous car liability project and Autonomous car liability project its completion, why is it that hardly any one uses it in information systems related Autonomous car liability projects?

1063. Where are your problem areas?

1064. When is it going to finish?

1065. Earned value can be used in almost any Autonomous car liability project situation and in almost any Autonomous car liability project environment. it may be used on large Autonomous car liability projects, medium sized Autonomous car liability projects, tiny Autonomous car liability projects (in cut-down form), complex and simple Autonomous car liability projects and in any market sector. some people, of course, know all about earned value, they have used it for years - but perhaps not as effectively as they could have?

1066. What is the unit of forecast value?

1067. Where is evidence-based earned value in your organization reported?

1068. How does this compare with other Autonomous car liability projects?

1069. Validation is a process of ensuring that the developed system will actually achieve the stakeholders desired outcomes; Are you building the right product? What do you validate?

1070. How much is it going to cost by the finish?

1071. Verification is a process of ensuring that the developed system satisfies the stakeholders agreements and specifications; Are you building the product right? What do you verify?

4.4 Risk Audit: Autonomous car liability

1072. Have permissions or required permits to use facilities managed by other parties been obtained?

1073. Are risk assessments documented?

1074. Are regular safety inspections made of buildings, grounds and equipment?

1075. Is all required equipment available?

1076. Is the customer technically sophisticated in the product area?

1077. Are you willing to seek legal advice when required?

1078. What are the benefits of a Enterprise wide approach to Risk Management?

1079. How do you compare to other jurisdictions when managing the risk of ....?

1080. How can the strategy fail/achieved?

1081. Are all managers or operators of the facility or equipment competent or qualified?

1082. Are procedures in place to ensure the security of staff and information and compliance with privacy legislation if applicable?

1083. Does your organization have a register of insurance policies detailing all current insurance policies?

1084. Do you conduct risk assessments on all programs, activities and events?

1085. Should additional substantive testing be conducted because of the risk audit results?

1086. Is the customer willing to establish rapid communication links with the developer?

1087. Have you considered the health and safety of everyone in your organization and do you meet work health and safety regulations?

1088. Does your organization have an up-to-date constitution?

1089. Does the Autonomous car liability project team have experience with the technology to be implemented?

1090. Does willful intent modify risk-based auditing?

4.5 Contractor Status Report: Autonomous car liability

1091. What are the minimum and optimal bandwidth requirements for the proposed solution?

1092. Who can list a Autonomous car liability project as organization experience, your organization or a previous employee of your organization?

1093. What process manages the contracts?

1094. If applicable; describe your standard schedule for new software version releases. Are new software version releases included in the standard maintenance plan?

1095. What was the overall budget or estimated cost?

1096. What was the actual budget or estimated cost for your organizations services?

1097. How does the proposed individual meet each requirement?

1098. How is risk transferred?

1099. Describe how often regular updates are made to the proposed solution. Are corresponding regular updates included in the standard maintenance plan?

1100. Are there contractual transfer concerns?

1101. How long have you been using the services?

1102. What was the budget or estimated cost for your organizations services?

1103. What is the average response time for answering a support call?

1104. What was the final actual cost?

4.6 Formal Acceptance: Autonomous car liability

1105. Do you perform formal acceptance or burn-in tests?

1106. Do you buy-in installation services?

1107. Was the client satisfied with the Autonomous car liability project results?

1108. Did the Autonomous car liability project manager and team act in a professional and ethical manner?

1109. What features, practices, and processes proved to be strengths or weaknesses?

1110. What was done right?

1111. Who supplies data?

1112. Did the Autonomous car liability project achieve its MOV?

1113. Does it do what Autonomous car liability project team said it would?

1114. Was the Autonomous car liability project goal achieved?

1115. Was the Autonomous car liability project work done on time, within budget, and according to specification?

1116. How does your team plan to obtain formal acceptance on your Autonomous car liability project?

1117. How well did the team follow the methodology?

1118. What is the Acceptance Management Process?

1119. General estimate of the costs and times to complete the Autonomous car liability project?

1120. What lessons were learned about your Autonomous car liability project management methodology?

1121. Do you buy pre-configured systems or build your own configuration?

1122. Does it do what client said it would?

1123. Was the Autonomous car liability project managed well?

1124. What are the requirements against which to test, Who will execute?

5.0 Closing Process Group: Autonomous car liability

1125. Were the outcomes different from the already stated planned?

1126. Did the delivered product meet the specified requirements and goals of the Autonomous car liability project?

1127. How well did the team follow the chosen processes?

1128. What do you need to do?

1129. How well did the chosen processes fit the needs of the Autonomous car liability project?

1130. Is the Autonomous car liability project funded?

1131. What is the overall risk of the Autonomous car liability project to your organization?

1132. What were things that you did very well and want to do the same again on the next Autonomous car liability project?

1133. What can you do better next time, and what specific actions can you take to improve?

1134. What is the risk of failure to your organization?

1135. Is this a follow-on to a previous Autonomous car

liability project?

1136. Just how important is your work to the overall success of the Autonomous car liability project?

1137. What areas does the group agree are the biggest success on the Autonomous car liability project?

1138. Are there funding or time constraints?

1139. How well did you do?

1140. What could be done to improve the process?

1141. Were escalated issues resolved promptly?

1142. What were things that you need to improve?

5.1 Procurement Audit: Autonomous car liability

1143. Was the submission of variant tenders accepted and duly ruled?

1144. Does the approval include approval of prices?

1145. Are goods generally ordered and received in time to be used in the programs for which they were ordered?

1146. Is the procurement function/unit organized the most appropriate way taking into consideration the actual tasks which the department has to carry out?

1147. Is there a purchasing policy as to the amount of an order on which bidding is required?

1148. Does the procurement function/unit understand costumer needs, supply markets and suppliers?

1149. Where your organization engaged an expert, was the contract awarded in compliance with procurement regulations?

1150. Are employees with cash disbursement responsibilities required to take scheduled vacations?

1151. Were exclusion causes duly considered before the actual evaluation of tenders?

1152. Does the procurement unit have sound commercial awareness and knowledge of suppliers and the market?

1153. Are fixed asset values recorded at historical cost?

1154. How is the evaluation of contract performance organized?

1155. Do buyers obtain price quotations or bids from two or more suppliers on significant purchases if catalog or advertised prices are not available?

1156. Are copies of policies made available to staff members involved in budget preparation and administration?

1157. Could bidders learn all relevant information straight from the tender documents?

1158. Are incentives to deliver on time and in quantity properly specified?

1159. Is the relationship between in-house and external work considered in the strategy?

1160. Was the decision on the award process accurate and adequately communicated?

1161. Are review meetings organized during contract execution and do they meet demand?

1162. Do procurement staff, supplier and end user communicate properly?

5.2 Contract Close-Out: Autonomous car liability

1163. Have all contracts been closed?

1164. Parties: Authorized?

1165. How does it work?

1166. Have all acceptance criteria been met prior to final payment to contractors?

1167. Change in circumstances?

1168. What is capture management?

1169. Has each contract been audited to verify acceptance and delivery?

1170. Was the contract type appropriate?

1171. Was the contract sufficiently clear so as not to result in numerous disputes and misunderstandings?

1172. Was the contract complete without requiring numerous changes and revisions?

1173. Parties: who is involved?

1174. What happens to the recipient of services?

1175. How is the contracting office notified of the automatic contract close-out?

1176. Change in attitude or behavior?

1177. Are the signers the authorized officials?

1178. How/when used ?

1179. Have all contract records been included in the Autonomous car liability project archives?

1180. Have all contracts been completed?

1181. Why Outsource?

1182. Change in knowledge?

5.3 Project or Phase Close-Out: Autonomous car liability

1183. What was expected from each stakeholder?

1184. What was the preferred delivery mechanism?

1185. What could have been improved?

1186. Were cost budgets met?

1187. What is a Risk Management Process?

1188. What are the mandatory communication needs for each stakeholder?

1189. What are the informational communication needs for each stakeholder?

1190. Did the delivered product meet the specified requirements and goals of the Autonomous car liability project?

1191. Was the schedule met?

1192. Complete yes or no?

1193. Does the lesson educate others to improve performance?

1194. What was learned?

1195. Is the lesson based on actual Autonomous

car liability project experience rather than on independent research?

1196. Is there a clear cause and effect between the activity and the lesson learned?

1197. Were risks identified and mitigated?

1198. Was the user/client satisfied with the end product?

1199. Have business partners been involved extensively, and what data was required for them?

1200. Planned completion date?

1201. Were messages directly related to the release strategy or phases of the Autonomous car liability project?

5.4 Lessons Learned: Autonomous car liability

1202. How effective was the quality assurance process?

1203. What solutions or recommendations can you offer that would have improved some aspect of the Autonomous car liability project?

1204. How efficient were Autonomous car liability project team meetings conducted?

1205. Will the information remain current?

1206. How many interest groups are stakeholders?

1207. What report generation capability is needed?

1208. What if anything has been lacking?

1209. Who had fiscal authority to manage the funding for the Autonomous car liability project, did that work?

1210. How effectively were issues resolved before escalation was necessary?

1211. How comprehensive was integration testing?

1212. What did you put in place to ensure success?

1213. Do you conduct the engineering tests?

1214. Did the delivered product meet the specified requirements and goals of the Autonomous car liability project?

1215. Why do you need to measure?

1216. Would you spend your own money to fix this issue?

1217. How useful was the content of the training you received in preparation for the use of the product/ service?

1218. What were the lessons learned on this Autonomous car liability project?

1219. What were the key issues?

1220. What data are likely to be missing?

1221. How effectively were issues managed on the Autonomous car liability project?

Index

abilities 197, 244
ability 32, 88, 196
absences 234
acceptable 50, 87, 144, 195
acceptance 8, 118, 186, 198, 253-254, 259
accepted 119, 186, 257
accepting 215-216
access 4, 9-11, 59, 144, 196, 234, 239
accomplish 9, 84, 105, 120, 132, 163, 195, 214, 243
according 30, 41, 157, 204, 253
account 37, 45, 246
accounted 177
accounting 245
accounts 193, 246
accuracy 153
accurate 11, 110, 163, 258
accurately 131, 238
achievable 118, 195
achieve 9, 77, 85, 112, 116, 130, 191-192, 204, 209, 218-219, 243, 248, 253
achieved 25, 82, 88, 124, 131, 249, 253
achieving 228
acquire 222
acquired 163
across 54, 141, 224
action 46, 96, 99, 101-102, 137, 203-204
actionable 54, 106
actioned 160
actions 27, 93, 98, 193, 195, 203, 232, 255
active 135, 152
actively 199
activities 24, 26-27, 41, 89, 97, 100, 109, 130, 160-165, 168, 174-175, 178, 180, 182, 202, 209, 229-230, 238, 250
activity 5-6, 31, 34, 156-157, 161, 163-165, 168-169, 173, 175, 177, 181-182, 232, 262
actors 135
actual 31, 44, 144, 156, 245, 251-252, 257, 261
actually 29, 60, 80, 99, 195, 201, 248
actuals 211
adapted 138

addition 123
additional 29, 62, 67, 71, 184, 232, 250
additions 101
address 1, 26, 75, 223
addressed 140, 169, 233
addressing 32, 105
adequate 28, 140-141, 160, 187, 198, 207, 211, 243
adequately 29, 177, 190, 230, 246, 258
adhered 157
Adjust 92, 100
adjusted 91, 228
adopted 137
adopters 200
advantage 1, 125
advantages 108
adverse 193
advertised 258
advice 249
advise 2
affect 58, 60, 66, 111-112, 119, 131, 148, 151, 186, 225-226, 233, 236-237
affected 132, 159, 199
affecting 13, 64
affordable 82
against 42, 94, 96, 142, 186, 211, 243, 254
agenda 131, 234
agendas 122
agents 187
aggregate 54
agreement 7, 121, 144, 198, 233
agreements 76, 197, 248
agrees 118
aiming 116
alerts 95
aligned 23
Alignment 216
alleged 3
alliance 86
allocate 107
allocated 45, 110, 157
allowable 46
allowances 234
allowed 2, 114, 171

allows 11, 167
almost 247
already 124, 193-194, 196, 221, 255
Although 130
altogether 206
always 11, 182
ambiguous 236
amount 160, 257
amplify72, 113
analysis 4, 8, 11-12, 66-70, 72, 89, 135, 171, 183, 189-190, 196, 223, 225, 234, 245
analyze 4, 58, 61, 71, 190, 208, 211
analyzed 95, 156, 207
another 154, 235
anothers 238
answer12-13, 17, 28, 44, 58, 74, 91, 104
answered 27, 43, 57, 73, 89, 103, 128
answering 12, 252
anybody 138
anyone 31, 127
anything 161, 169, 173, 186, 189, 227-228, 263
appear 3
applicable 12, 92, 157, 190, 228, 249, 251
applied 78, 97, 229
appointed 36, 41
appraise 135, 223
approach 79, 87, 108, 122, 141, 221, 235, 243-244, 249
approaches 74, 88, 138, 187, 199
approval 31, 111, 142, 257
approvals 187
approve 144, 224
approved 38, 73, 148, 150, 174, 179, 187, 225
approving 148
architect 151
Architects 9
archives 260
around117, 127, 194, 239
articulate 235-236, 243
artifact236
asking 3, 9, 185, 191
aspect 263
aspects 213, 243
assess 22, 40, 81, 95, 111, 167, 206

assessed 81, 160, 203-204
assessing 75, 92
Assessment 7-8, 10-11, 18, 145, 205, 235, 237-238
assets 47, 135
assign 25
assigned 141, 151, 154, 156, 164, 178, 180, 193, 197, 211, 217, 231-232, 246
assignment 6, 163, 193-194, 246
assist 10, 72, 93, 184, 219
assistant 9
associated 146
assume 198
assuming 206
Assumption 5, 152
assurance 25, 141, 159-160, 179, 212, 228, 263
attainable 35, 195
attempted 31
attempting 96
attend 19
attendance 41
attendant 77
attended 1, 41
attending 201
attention 13, 112
attitude 260
attitudes 221
attributes 5, 109, 146, 163, 224
audience 218
audited 259
auditing 26, 97, 105, 245, 250
audits 229
author 3
authority 67, 130, 144, 156-157, 196, 199, 229, 236, 263
authorized 139, 156, 193, 245-246, 259-260
automatic 259
Autonomous 3-15, 17-27, 29-44, 46-63, 65-142, 144-148, 150-152, 154-157, 159-169, 171, 173-181, 183-187, 189, 191, 193-209, 211-213, 215, 217, 219-227, 229, 231-233, 235, 237, 239, 241-243, 245-251, 253-257, 259-264
available 19, 29, 48, 60, 69, 77, 93, 109, 158, 161, 163-164, 174, 177, 180, 190, 193, 202, 205-206, 227, 249, 258
Average 13, 27, 43, 57, 73, 89, 103, 128, 176, 197, 228, 252
avoiding 201

awarded 257
awareness 258
background 11, 165
backup 236
backward 225
balanced 82
bandwidth 251
barriers 118, 199
baseline 6, 86, 140, 158, 185
baselined 198, 212
baselines 33-34
basics 113
because 2, 178, 250
become 105, 117, 119, 125, 144, 148, 228
before 1-2, 11, 31, 98, 151, 168, 175, 224, 257, 263
beginning 4, 16, 27, 43, 57, 73, 90, 103, 128
behavior 173, 244, 260
behaviors 22, 195
behaviour 230
behaviours 217
behind 2
belief 12, 17, 28, 44, 58, 74, 91, 104, 122
beliefs 237
believable 118
believe 2, 118, 122
benefit 3, 19, 21, 25, 54, 94, 203
benefits 26, 44, 48, 51, 69, 71, 104, 106, 114, 124, 126, 167, 236-237, 249
besides 173
better 9, 36, 53, 79, 135, 161, 175, 184, 210, 227-228, 232-233, 236, 255
between 72, 130, 148, 153, 156, 172-173, 188, 199, 216, 224, 244, 246, 258, 262
beyond 235
bidders 145, 258
bidding 257
biggest 57, 80, 219, 256
blinding 60
bother 50
bothered 239
bottleneck 162
bounce 61, 63
boundaries 32, 141

bounds 32
Breakdown 5, 154, 171, 173
briefed 37
brings 39
broken 61
budget 2, 91, 94, 109, 140, 142, 157-158, 161, 178, 185, 193, 212, 245-246, 251-253, 258
budgeted 44, 156, 245
budgets 25, 115, 156-157, 194, 196, 246, 261
building 27, 96, 138, 153, 171, 174, 248
buildings 249
burden 228
burn-in 253
business 2, 9, 11, 21, 24, 35-36, 49, 52, 54, 69-70, 78, 88, 98, 112-113, 115-117, 124, 127, 130, 133, 142, 145-146, 151, 157, 179, 207, 217-218, 225, 262
busywork 131, 219
buyers 258
buy-in 107, 253
calculate 165, 174
called 137
capability 22, 132, 234, 263
capable 9, 43
capacities 123
capacity 22, 27, 88, 132
capital 120
capitalize 65, 243
capture 54, 98, 153, 190, 259
captured 54, 59, 76, 111, 160, 211
capturing 227
career 148, 195
careers 121
carried 65
carrying 224
cash-flow 135
catalog 258
catching 1
category 34, 156
caused 3, 49, 54
causes 46, 56-58, 61-62, 68, 98, 140, 157, 193, 245, 257
causing 153
celebrate 76
center 140

central 227
centrally 82
certain 200
Certified 159, 197
challenge 9, 227-228
challenges 114, 165
change 7, 17-18, 34, 45-46, 60, 64, 68-70, 80, 87, 89, 100, 110, 132, 134, 139-141, 144, 150, 155, 164, 173, 178, 182, 186, 189, 197, 208, 217-218, 223-226, 239, 241, 259-260
changed 20, 34, 82, 116, 177, 182-183, 203, 223
changes 19, 32, 34, 48, 73, 77, 81, 87, 98, 101, 118, 120, 122, 132, 139, 142, 144, 150-151, 153, 156-157, 171, 174, 186, 194-195, 205, 207, 218, 223, 225, 237-238, 245, 259
changing 92, 127
channel 130
channels 146, 220
charge 124
chargeable 245
charged 55, 194
charter 4, 39, 41, 77, 132-133, 139, 159, 197-198, 215
charts 71, 177
cheaper 53
checked 65, 93, 96-97
checklists 10
choice 34, 123
choose 12, 87
chosen 131, 137, 206, 255
circumvent 22
claimed 3
clarify 121
clarity 151
classes 141, 173, 179, 211
classified 157
clearly 12, 17, 20, 28, 30-31, 37, 44, 58, 70, 74, 86, 91, 104, 142, 196, 215, 244
client 50, 127, 130, 209, 253-254, 262
clients 20, 34
closed 102, 225, 259
closely 11
Close-Out 8, 213, 259, 261
closest 119
Closing 8, 71, 255
coaches 30, 191-192

coherently 229
colleague 114
colleagues 108, 175
collect 66, 99, 183, 187, 191, 232
collected 36, 40, 59, 61-62, 64, 68
collection 66, 191
combine 88
coming 71, 214
command 97
commercial 258
commit 197, 235
commitment 93, 119, 136, 216
committed 60, 142, 160, 179, 201-202, 211
Committee 152, 216
common 132, 178, 181, 214
community 176, 183, 209
companies 3, 93, 227
company 1-2, 9, 53, 110, 112-114, 117, 126
comparable 139
compare 69, 82, 203, 245, 248-249
compared 124, 157, 197, 211
comparing 74, 139
comparison 12
compatible 225, 243
compelling 35
competent 249
competing 56
competitor 2
compilers 202
complete 3, 10, 12, 26, 32, 37-38, 140, 159, 163-164, 172-173, 176, 189, 195, 198, 207, 233, 254, 259, 261
completed 13, 33, 35, 40, 151, 157, 168, 174, 232, 260
completely 1, 127, 175
completing 110
completion 34, 39, 131, 156-157, 167, 244, 246-247, 262
complex 9, 119, 247
complexity 20, 49
compliance 1, 19, 51, 71, 76, 153, 249, 257
complied 232
components 141, 152, 156, 174, 177
compute 13
computing 127
concept 84, 166, 191

concern 48, 86, 196
concerned 18
concerns 2, 20, 23, 119, 199, 251
concrete 76, 244
condition 93, 137
conditions 97, 108, 186, 232
conduct 214, 250, 263
conducted 152, 198, 214, 225, 250, 263
conference 214
confidence 189, 201
confirm 12, 173
conflict233
conflicts 219, 232
conform 187
conjure 147
connected 178
connecting 114
consensus 234
consider 22, 24, 132, 163, 214
considered 19, 24, 52, 250, 257-258
consist 246
consistent 34, 64, 98, 157, 201
constantly 1, 239
Constraint 5, 152
consult 1
consultant 1-2, 9, 181, 205
consulted 116
consulting 2, 53
consumers 115
contact 9, 180, 215
contain 102
contained 3
contains 10
content 35, 140, 224, 264
contents 3-4, 10
context 34-36, 219
continual 100, 102
continuity 52
continuous 71, 78
contract 8, 157, 159, 182, 193, 214, 222, 231-232, 245, 257-260
Contractor 8, 145, 205, 251
contracts 31, 136, 251, 259-260

contribute 147, 230
control 4, 30, 52, 71, 91, 93-94, 97, 99-102, 131, 143, 156, 159, 174, 183, 193, 223-224, 228, 237, 246
controlled 66, 157, 203, 239
controls 59, 70, 76, 89, 92, 94-95, 97-98, 101, 210
convention 123
converge 243
convey 3
cooperate 183
copies 258
Copyright 3
corporate 2, 229
correct 44, 91, 177, 211
corrective 98, 203
correctly 138
correspond 10-11, 156
costing 52
costumer 257
counsel 213
counting 115, 145
counts 115
course 34, 46, 247
covered 131
covering 10, 102
coworker 127
crashes 49
crashing 167
craziest 121
create 24, 66, 107-108, 113
created 65, 72, 134, 139, 173-174, 180, 209
creating 9, 57, 141, 153
creative 18
creativity 79
credible 183
crisis 18
criteria 4, 7, 10-11, 34-35, 41, 64, 78, 81, 92, 107, 129-131, 149, 174, 183, 185, 189, 213, 259
CRITERION 4, 17, 28, 44, 58, 74, 91, 104
critical 35-36, 42, 72, 81, 94, 101, 122, 158, 162, 165, 219, 227, 237
criticism 65
criticized 235
cross-sell 113

crucial 60
crystal 12
culture 32, 64, 195-196, 217
cultures 233
cumbersome 202
current 40, 44, 48, 52, 63, 69, 72, 82, 89, 91, 118, 122-123, 147, 153, 192, 195-196, 198, 217, 223, 250, 263
currently 41, 126, 227
custom 21
customer 24, 28, 30, 34, 36, 40, 42, 85, 96, 100, 107, 114, 120, 126, 146, 150, 185-186, 201, 232, 249-250
customers 3, 21, 29, 33, 53-56, 63, 69, 81, 106, 108-109, 111, 113, 117-119, 123-125, 132, 146, 165, 205, 210, 222, 231, 245
customized 2
cut-down 247
damage 3
Dashboard 10
dashboards 101
databases 191
day-to-day 100, 114
deadlines 19, 106, 136
dealer 159
dealing 21
debriefing 214
deceitful 127
decide 87, 196
decided 87
deciding 124
decision 7, 48, 63, 81, 83-84, 87, 227-228, 234, 258
decisions 75-77, 79-80, 82, 84, 87, 93, 96, 99, 136, 196, 211, 220, 227, 231-232, 236
decomposed 174
decrease 190
dedicated 9
deeper 12
defect 141, 179, 211
defects 187, 189
define 4, 28, 30, 32, 38, 61, 67, 79, 132, 145-146, 154, 175, 189, 227, 235
defined 12, 17-18, 28, 30-34, 37, 40-41, 44, 58, 62, 66, 74, 91, 104, 141, 150-151, 154, 173, 177-178, 190, 194, 196, 211-212, 215, 220
defines 26, 31, 37

defining 9, 110
definite 102
definition 20, 23, 29, 33, 36, 39, 152, 180, 185
degree 201, 235-238, 243-244
-degree 2
delayed 165
delaying 46
delays 56, 176-177
delegated 42
delete 182
deletions 101
deliver 21, 42, 85, 120, 124, 184, 229, 234, 258
delivered 47, 145, 171, 232, 242, 255, 261, 264
delivering 237
delivery 22, 50, 108, 123, 259, 261
demand 81, 111, 118, 258
demands 197, 243-244
department 9, 113, 156, 193, 198, 222, 257
depend 221
dependent 122
depends 112
depict 167
deploy 102
deployed 97
deploying 53
deployment 52
derive 92
Describe 21, 142, 148, 153, 166, 251
described 3, 150, 225, 246
describing 33
deserving 222
design 11, 61, 67, 80, 89, 97, 122, 137
designated 159, 179
designed 9, 11, 67, 87
designing 9
desired22, 29, 64, 77, 171, 173, 225, 237, 248
detail 83, 132, 178
detailed 63, 72, 152, 160, 194, 198, 215, 238, 246
detailing 250
details 54, 218
detect 97
determine 11, 106, 112, 183, 213, 233
determined 59, 112, 132, 191

detracting 124
develop 74, 83-84, 86, 88, 154, 235
developed 11, 39-40, 42, 54, 77, 139, 152, 160-161, 166, 179, 191, 212, 217, 225, 238, 248
developer 250
developers 202
developing 72, 83, 215, 237
develops 237
device 107
diagram 5, 48, 54, 61, 167
diagrams 50
dictates 178
Dictionary 5, 156
differ 177
difference 172-173, 181, 227
different 9, 27, 29-30, 37, 61, 72, 112, 123, 138, 188, 195, 197, 208, 220, 227, 255
difficult 67, 162, 164, 168, 181
dilemma 108
direct 156, 245
direction 34, 53, 139-140
directions 209
directly 3, 63, 69, 130, 136, 262
Directory 7, 231-232
Disagree 12, 17, 28, 44, 58, 74, 91, 104
disaster 46, 52, 178
disclosure 93, 157
discounts 81
discover 145
discrete 246
discussion 122
display 165
displayed 36, 64, 161, 175
disposed 230
disputes 259
disqualify 60
disruptive 70
distribute 231
Divided 27, 42-43, 57, 73, 89, 102, 128
document 11, 131, 144-145, 152-153, 227, 232
documented 40, 86, 88, 92, 96, 102, 147, 152-153, 159, 187, 189, 196-198, 212, 215, 220, 224, 249
documents 9, 174, 188-189, 258

domains 85
dominating 233
drawbacks 193
Driver 70
drivers 48, 69
drives 45
driving 107
duration 6, 137-138, 150, 155, 168, 173, 175
durations 31, 177
during 34, 85, 163, 169, 207, 209, 214, 227, 258
duties 224, 238
dynamics 39
eagerly 2
earlier 114
earned 8, 247
earning 174
easily 235
economic 209
economical 139
economy 88
eDiscovery 228
edition 10
editorial 3
educate 113, 261
educated 1
education 24, 92
effect 138, 205, 262
effective 18-19, 105, 115, 147, 175, 201, 227, 229-230, 263
effects 46, 165, 231
efficiency 63, 95, 214
efficient 50, 86, 137, 160, 221, 263
effort 37, 46-47, 109, 141, 156, 230, 237, 239
efforts 31, 85, 132
either 163
Electrical 152
electronic 3
elements 11, 35, 72, 99, 112, 156-157, 159, 191, 231, 246
Elevator 146
elicit 192
e-mail 146
embarking 35
embrace 217
emerging 1, 66, 96, 237

emissions 111
employee 87, 104, 251
employees 21-22, 25, 112-113, 118, 188, 257
employers 134
empower 9
enable 70
enablers 107
encourage 79, 101, 214
encouraged 193
end-users 202, 206
energy 1, 111
engage 123, 216, 235, 239, 244
engaged 257
engagement 54, 134
Engineers 152
enhance 98, 243
enhanced 121
enhancing 97
enough 9, 73, 111-112, 119, 149, 180, 243
ensure 31, 42, 61, 77, 105, 110, 116, 118, 138, 174, 187-188, 191, 195-196, 198, 219, 232, 236, 249, 263
ensures 116
ensuring 11, 110, 230, 248
Enterprise 221, 249
entire 193
entities 55
entitled 214
entity 3, 237
equipment 22, 26, 174, 205, 234, 249
equitably 42, 236
errors 119, 187
escalated 256
escalation 185, 263
especially 174
essential 88, 236
essentials 111
establish 74, 102, 183, 250
estimate 49-50, 52, 139-140, 150, 176, 178, 201, 213-214, 254
estimated 34, 39, 48, 52, 120, 141, 183-184, 201, 251-252
estimates 6, 39, 50, 59, 139-140, 152, 156-157, 173, 178, 181, 186, 193, 211, 246
estimating 6, 160, 175, 181, 183-184, 198, 215

estimation 86, 142, 211
estimator 181
estimators 181
etcetera 49, 108
ethical 110, 253
ethics 207
ethnic 113
evaluate 75, 80, 84, 194, 207, 215
evaluated 131, 186
evaluating 76, 78, 86
evaluation 64, 82, 88, 99, 213, 244, 257-258
evaluators 240
events 19, 76, 79, 83, 246, 250
Everyday 1
everyone 36, 43, 150, 229, 250
everything 51
evidence 12, 48, 203, 221
evolution 44
evolve 99
exactly 189
examined 36
example 4, 10, 14, 22, 63, 98, 140-141, 152, 156-157
examples 9-10, 133
exceed 155, 168
exceeding 54, 195
excellence 9, 36
excellent 57
excelling 195
excess 157
exchange 244
excited 1
exclude 87
excluded 158
exclusion 257
Exclusions 180
execute 254
executed 186, 226
Executing 7, 207, 219
execution 102, 153, 258
executive 9, 107
executives 122
Exercise 25, 168
existence 210

existing 11, 98, 113, 132, 145, 151
exists 144
expect 105
expected 26, 31, 87, 122, 124, 137, 140, 174, 182, 229-230, 261
expend 47
expense 156
expenses 233
experience 41, 119, 122, 126, 165, 175, 181, 194, 245, 250-251, 262
experiment 108
expert 257
expertise 88, 140, 193, 206
experts 31, 208
explained 11
explicitly 115
explore 61, 243
exposures 86
extensive 2
extent 12, 18, 20-21, 36, 79, 137-138, 193, 223, 230
external 1-2, 31, 113, 157, 160, 190, 206, 258
facilitate 12, 22, 68, 101, 244
facilities 206, 249
facility 249
facing 22, 108, 205
fact-based 235
factored 135
factors 46, 78, 124, 132, 138, 168, 184
failed 49
failing 204
failure 55, 110, 126, 173, 190, 255
fairly 42
familiar 10
fashion 3
favorable 193, 245
feasible 50, 109, 159, 183
feature 11
features 253
feedback 34, 40, 49, 131, 233
feeling 1
fields 135
finalize 213
finalized 14

financial 50, 59, 64, 69, 120, 137
fingertips 11
finish 133, 165, 247-248
finished 133, 165
fiscal 263
follow 102, 106, 137, 167-168, 254-255
followed 40, 142, 179, 223
following 10, 12, 188
follow-on 255
for--and 94
forces 189
forecast 198, 247
forefront 122
foresee 165
forever 116
forget 11
formal 8, 119, 142, 144, 152, 179, 197, 253-254
formally 32, 157, 197, 201
format 11, 224
formats 202, 237
formula 13, 123
Formulate 28
formulated 234
forward 2, 117
forwarded 156, 246
foster 109
framework 97, 126, 191
freaky 111
frequency 41, 97, 105, 230
frequent 179
frequently 44, 235, 243
friction 236
friend 107-108, 114
friends 2, 236
frontiers 87
fulfill 125
full-blown 48
full-scale 89
function 135, 141, 151, 257
functional 139
functions 32, 58, 117, 126, 141, 146, 223, 231
funded255
funding 116, 120, 152, 243, 256, 263

further 10
future 9, 56, 94-95, 122
gained 2, 71, 97, 100
garbage 230
gather 12, 28-30, 32, 35, 38, 41, 44, 62, 65-66
gathered 33, 59, 62, 67, 70, 73
gathering 36-37, 42, 147
general 80, 132, 198, 254
generally 221, 257
generate 67-68, 190
generated 63, 157
generation 10, 68, 263
generators 202
generic 2
getting 2, 49
global 88, 127, 166
govern 125, 244
governance 24, 114, 187, 212, 238
governing 230
government 174
granted 234
graphical 177
graphics 27
graphs 10
greater 140
greatest 74, 230
ground 69
grounds 249
groups 112, 159-160, 171, 217-218, 222, 244, 263
growing 173
growth 60, 124
handle 169, 206, 209, 234
handled 157, 213, 234
happen 21, 106, 204, 209
happened 131
happening 127
happens 9, 34, 54, 104, 119, 121, 160, 183, 209, 259
hardest 52
hardly 247
hardware 201, 206
havent 117
having 234, 245
hazards 209

health 111, 250
healthy 188
hearing 113
helpdesk 224
helping 9, 132, 219
hidden 51
highest 21
high-level 33, 35, 146
highlight 2, 195
Highly 70
high-tech 105
hijacking 112
hiring 101
historical 183, 245, 258
history 162, 181, 185
hitters 71
hitting 247
holders 130, 142
holiday 2
honest 110
Honestly 2
hoping 132
horizon 118
housed 144
humans 9
hypotheses 58
identified 3, 19, 24-25, 28, 30, 70-72, 75, 137, 141, 151-153, 156-158, 164, 185, 193-194, 196-198, 204, 209, 216, 232-233, 262
identifies 207
identify 1, 11-12, 20-21, 24, 63, 66, 68, 86, 160, 165, 174, 183, 193, 217, 245
ignore 18
ignoring 122
images 147
imbedded 100
impact 7, 40, 46, 49-50, 52, 56-57, 78, 108, 131, 150, 194, 203-205, 207, 225, 238, 241-242
impacted 46, 217-218
impacts 52, 55, 136, 175, 201, 203, 215
implement 27, 47, 61, 91
implicit 120
importance 237

important 18, 24, 34, 63-64, 69, 110, 115, 117, 121, 126, 130, 136, 150, 199, 234, 236, 256
improve 4, 11, 61, 74-75, 79-83, 85-88, 132, 152, 174, 182, 232, 255-256, 261
improved 2, 81, 83, 93, 261, 263
improving 84, 221
inactive 182
inadequate 1
incentives 101, 258
incident 227
include 24, 65, 87-88, 139, 151, 158-159, 161, 174, 213, 223, 257
included 4, 9, 18, 51, 146, 150, 175, 183, 208, 251, 260
INCLUDES 11
including 22, 30, 36, 41, 50, 53, 62, 96, 99, 193, 246
increase 79, 110
increased 118, 195
increasing 117
incurring 156, 194
in-depth 10, 12
indicate 60, 93, 230
indicated 98
indicators 24, 47, 55, 63, 69, 71-72, 89, 96, 194
indirect 55, 156, 193-194
indirectly 3
individual 45, 178, 193, 211, 233, 244, 251
industry 1-2, 121, 124
infinite 127
influence 87, 115, 130, 134, 196, 228
influences 166, 200, 233
informal 243-244
informed 117
ingrained 95
inherent 111
in-house 2, 258
initial 29, 124
initially 35
initiated 151, 184, 189, 226
Initiating 4, 118, 130
initiative 12, 192, 199, 239
Innovate 74
Innovation 63, 88, 99, 122
innovative 108, 137, 184, 241

in-process 71
inputs 33, 37, 54, 59-60, 94, 133, 173
inside 21
insight 67, 69
insights 1-2, 10
inspection 142, 215
inspired 106
Instead2
Institute 152
insurance 173, 250
insure 117
integrate 78, 99, 124
Integrated 186
integrity 21, 106, 225
intended 3, 77
INTENT 17, 28, 44, 58, 74, 91, 104, 250
intention 3
interact 126
interest 119, 174, 222, 263
interests 199
interface 201, 206
interim 106
internal 1, 3, 31, 69, 111, 113, 142, 156, 166, 190, 211, 246
interpret 12
intervals 142, 211
interview 1, 109
introduce 208
invest 71
investing 2
investment 18, 56, 66, 192
investor 46
involve 113
involved 18, 21, 40, 58, 83, 125, 144, 147, 160, 198, 215, 222, 236-237, 241, 258-259, 262
involves 92
isolate 193, 245
issues 18-21, 23, 132, 136, 146, 150-151, 160, 167, 169, 179, 193, 219, 229, 256, 263-264
iteration 141
itself 3, 23
justified95, 130
killer 108
know-how 206

knowledge 1-2, 11, 31, 38, 41, 71, 81, 88, 94, 97-101, 107, 109, 126, 157, 217, 258, 260
lacking 263
languages 233
largely 72
latest 10, 157
leader 19, 63-64, 89
leaders 36, 59, 61, 93, 142, 179, 200, 211
leadership 25, 38, 40, 88, 111, 136
learned 1, 8, 98, 111, 191, 198, 254, 261-264
learning 94, 98-99, 227
lesson 261-262
lessons 8, 89, 98, 111, 191, 198, 254, 263-264
letter 146
Leveling 164
levels 21-22, 31, 63, 69, 89, 94, 111, 154, 157, 210
leverage 43, 88, 98, 119, 183
leveraged 31
leveraging 237
liability 3-8, 10-15, 17-27, 29-44, 46-63, 65-142, 144-148, 150-152, 154-157, 159-169, 171, 173-181, 183-187, 189, 191, 193-209, 211-213, 215, 217, 219-227, 229, 231-233, 235, 237, 239, 241-243, 245-251, 253-257, 259-264
licensed 3
licenses 158
lifecycle 52, 63
lifecycles 88
Lifetime 11
likelihood 77, 79, 202, 204
likely 84, 91, 114, 214, 227-228, 264
limitation 57
limited 11, 196
linked 30
listed 180, 191
listen 106, 125
little 2
locally 82
location 135, 227
logged 197, 223
logical 169
logically 160
longer 1, 97
long-term 102, 114, 123

looked 1
losing 55
losses 22, 29
magnitude 76
maintain 91, 106, 124, 188
maintained 75
majority 234
makers 83, 98
making 19, 63, 87, 191, 219, 234
manage 33-34, 42, 45, 51, 53, 61, 77, 79, 83-85, 107, 115, 130, 139, 148, 159, 178, 181, 191, 198-199, 215, 263
manageable 38, 83, 141, 174
managed 9, 33, 67-69, 76, 78, 81, 83, 100, 197, 249, 254, 264
management 5-7, 10-11, 18-19, 24, 30, 41, 51, 59, 64, 68-69, 71-72, 78-79, 81-82, 84, 86, 107, 111, 115, 119, 139-141, 144, 151, 156, 158-160, 171-172, 174, 178-179, 185, 187-188, 193, 197-202, 204-206, 211, 213, 215, 217, 219-221, 223, 228, 231, 241-242, 247, 249, 254, 259, 261
manager 9, 11, 33, 42, 130, 141, 173, 197, 205, 213, 222, 253
managers 4, 129-130, 201, 219-220, 249
manages 76, 84, 139, 251
managing 4, 88, 129-130, 134, 178, 208, 229, 249
mandatory 225, 261
manner 22, 80, 157, 160, 220, 223, 244, 246, 253
manpower 156
Manual 229
Mapping 65, 67
margin 159
market 17, 136, 201, 247, 258
marketable 201
marketer 9
marketing 115, 146
markets 26, 257
material 132, 156
materials 3, 229
matrices 148
Matrix 4-7, 135, 148, 189, 193, 207
matter 31, 47, 55
matters 228
maximize 237
maximizing 104
maximum 130

meaning 164
meaningful 57, 114, 194
measurable 35, 42, 131, 236
measure 4, 11, 23, 26, 32, 43-46, 49-52, 57, 62, 71, 74, 77, 80, 94-96, 98-99, 135, 183, 185-186, 190, 207, 243, 264
measured 17, 46-48, 53, 55-56, 77, 94, 96, 209
measures 44-45, 50, 53-56, 63, 68-69, 71, 89, 93, 96, 137, 203, 224, 241
measuring 97, 158, 194
mechanical 3
mechanism 153, 208, 223, 261
mechanisms 135, 150
medium 247
meeting 33, 38, 96, 131, 142, 196, 222, 228, 231, 233-234, 239
meetings 39, 41, 131, 142, 201, 211, 233-234, 258, 263
member 7-8, 40, 111-112, 172, 199, 207, 221, 233, 237
members 1, 30, 43, 93, 142, 144, 160, 180, 198, 200, 206, 215, 231, 233, 235-238, 243-244, 258
membership 235, 243
memory 223
mentors 205
message 92
messages 199, 233, 262
method 47, 163, 189, 234-236
methods 34, 41, 50, 62, 184, 187-188, 207, 230, 233
metrics 6, 31, 66, 76, 101, 189-190, 212, 215
Microsoft 219
milestone 5, 141, 165-166, 197
milestones 41, 134, 159, 167
minimize 131, 187, 241
minimizing 63, 104
minimum 251
minutes 38, 81, 231
missed 52, 113
missing 68, 118, 163, 185, 264
mission 60, 125-126, 132, 196
Mitigate 79, 131, 241
mitigated 2, 262
mitigating 216
mitigation 135, 139, 197, 208
mobile 237
Modeling 72, 192

models 17, 51, 70, 116, 153
modern 211
modified 102
modify 250
module 144
moment 108
moments 60
momentum 113, 122
Monday 1
monetary 19
monitor 93-94, 96, 99-100, 183, 197
monitored 93, 95, 100, 161
monitoring 8, 95-97, 100, 169, 205, 241
monthly 245
months 1, 80-81
Morale 136
morning 1
motivate 119, 244
motivation 20
motive 191
moving 117
multiple 240
narrow 62
national 136, 209, 233
native 233
nearest 13
nearly 118
necessary 61-62, 70, 87, 108, 116, 122, 183, 187, 218-219, 226, 229, 235, 263
needed 2, 21-22, 24, 26-27, 37, 59-60, 70, 92, 95, 98, 102, 142, 152, 176, 199, 218, 220, 263
negative 177
negotiate 127
negotiated 121
neither 3
Network 5, 167
Neutral 12, 17, 28, 44, 58, 74, 91, 104
normal 95
notice 3, 138
notified 218, 259
noting 223
number 27, 43, 56-57, 73, 89, 102, 128, 163, 171, 174, 207, 265

numbers 119
numerous 259
objection 18, 20
objective 9, 50, 135, 182, 186
objectives 2, 23-25, 28, 30, 40, 60, 92, 99, 106, 120-121, 147, 174, 188, 209, 212, 216, 222, 236
observe 195
observed 75
observing 153
obstacles 22, 166, 184
obtain 106, 254, 258
obtained 40, 187, 249
obtaining 54
obviously 12
occurring 88, 207
occurs 18, 46, 98, 190, 241
offerings 69, 82
office 159, 188, 214, 221, 259
officer 1, 213
officials260
offsite 208
onboarding 215
one-time 9
ongoing 86, 94, 161
on-going 197
on-site 232
operate 209, 223
operates 121
operating 7, 49, 53, 98, 194, 233
operation 101, 175
operations 11, 95, 97, 99-100
operators 102, 189, 249
opponent 227
opponents 240
opposed 199
opposite 121-122
opposition 117
optimal 81, 251
optimize 87, 97
optimized 106
option 123, 237
ordered 1, 257
organize 140, 173

organized 257-258
orient 96
original 178
originate 99
others 178, 180, 183, 186, 194, 199, 201, 223, 227, 235, 261
otherwise 3, 157
outcome 12, 132, 171, 237
outcomes 85, 97, 113, 184, 248, 255
outlier 163
outlined 92
output 43, 60, 64, 66-68, 70, 73, 93, 97
outputs 33, 59, 64, 66-67, 94, 169, 173
outside 79, 220, 233-234
outsource 65, 173, 260
outweigh 48
overall 11-12, 23, 52, 99, 111, 116, 130, 147, 167, 177, 205, 228, 251, 255-256
overcome 184
overhead 156-157, 193-194, 245-246
overheads 179
overlook 239
overseas 205
oversight 68, 152, 159, 237
overtime 170
owners 154
ownership 30, 99
package 246
packages 194, 246
Padding 181
paradigms 123
parallel 167
parameters 101
Pareto 71
parking 234
particular 63, 206
parties 2, 86, 249, 259
partners 21, 40, 86, 91, 112, 118, 136, 205, 236, 262
patterns 88
paying 112
payment 179, 259
pending 225

people 9, 25, 57, 65, 72, 83, 96, 101, 108, 110, 112, 119-121, 125-126, 135, 153, 174, 191, 193-194, 198-199, 201, 207, 217, 220, 222, 247
perceive 125
perceived 236
percent 114
percentage 148
perception 86, 110
perform 25, 32, 43, 223, 236, 253
performed 86, 148, 161-162, 180, 201, 204, 215, 245
performers 197
performing 131
perhaps 20, 247
period 87, 203
periodic 179
permission 3
permits 249
person 3, 165, 194
personal 115
personally 148
personnel 19, 95, 180, 192, 230
pertinent 95
phases 52, 145, 262
pitfalls 111
planet 96
planned 92, 100-101, 144, 160-161, 193, 203, 255, 262
planners 99
planning 4, 10, 93, 98, 137, 151-152, 156, 161, 169, 201, 222, 241
platform 237
players 85
playing 2
pocket 176
pockets 176
points 27, 43, 57, 64, 73, 89, 102, 127-128, 166, 191
policies 195, 229, 250, 258
policy 40, 79, 98, 169, 187, 229, 257
Political 36, 127, 137, 165
portfolio 111, 241
portfolios 199, 239
portion 2, 177
portray 71
position 195-196

positioned 183-184
positive 80, 113, 139-140, 153, 240
possess 243
possible 49, 57, 62, 68, 79, 91, 123, 127, 141, 175, 179, 205, 211, 235
potential 24, 46, 60, 75, 88, 113, 157, 199, 204, 233, 237
practical 74, 83, 91, 221-222
practice 197
practices 11, 63, 88, 98, 101, 211, 237, 253
precaution 3
precede 168
predicting 97
prediction 163
predictive 171
predictor 171
preferred 261
pre-filled 10
prepare 195, 214
prepared 1
present 94, 122, 127, 135, 159, 217
presented 1, 20, 142
preserve 41
pressures 136
prevalent 218
prevent 51, 219, 230, 233
prevents 19
previous 31, 166, 251, 255
previously 225
prices 257-258
primary 47, 171
principles 211
priorities 47, 50-51, 54, 56, 186
prioritize 208
priority 46, 49, 164, 217, 237
privacy 39, 190, 249
probably 207
problem 17, 19-20, 22-24, 28, 31, 35-36, 41, 49, 56, 71, 147, 221, 228, 235, 247
problems 18, 22-23, 84, 86, 88, 98, 111, 135, 137, 146-147, 191, 241
procedure 157, 159, 229
procedures 11, 88, 92, 98, 102, 141-142, 158, 169, 177, 179, 188, 190, 198, 223, 228, 249

proceed 206, 245
proceeding 175
process 4, 6-9, 11, 29, 33, 35, 37, 40, 42-43, 48, 59-66, 68-70, 72-73, 75, 80, 91-97, 99-101, 130, 137, 145-148, 150, 152-153, 159, 169, 181, 185, 187, 191, 197, 201-202, 204-206, 213, 215, 219, 227, 231-232, 234, 236-237, 241, 248, 251, 254-256, 258, 261, 263
processes 1, 52, 58, 60-62, 65, 67-71, 98-100, 131, 137, 141, 192, 202, 218, 220, 225, 229, 253, 255
procuring 157
produce 1, 67, 133, 169, 222, 231
produced 80, 150, 215
produces 163
producing 148
product 3, 52, 63, 65, 69, 108, 122, 132, 151, 166, 186, 201-202, 209, 221-222, 226, 232, 248-249, 255, 261-262, 264
production 86, 118
products 3, 22, 26, 57, 113, 116, 137-138, 148, 201, 206, 222, 225
profile 81
profit 190
program 18, 55, 59, 94, 138, 181, 187, 241
programme 219
programs 199, 222, 239, 250, 257
progress 37, 53, 80, 99, 112, 138, 183, 191, 198, 219, 243
prohibited 157
project 4-6, 8-10, 21, 25, 40, 48, 58, 60, 63, 81, 92, 94, 100, 108, 113, 115-117, 119-120, 123, 126-127, 129-134, 136-142, 144-145, 147-148, 150-152, 154-155, 159-162, 164, 166-168, 171-181, 183-187, 189, 194, 196-202, 204-207, 211-213, 215, 217, 219-226, 231-232, 241-243, 246-247, 250-251, 253-256, 260-264
projected 157
projects 4, 51, 114, 127, 129, 137, 148, 150, 154, 185-186, 196, 199, 208, 211, 219, 221-222, 233, 239, 247-248
promising 108
promote 57, 65, 173
promotion 195
promptly 256
proofing 75
proper 93
properly 31, 36, 157, 229, 258
property 229
proponents 240
Proposal 166, 213

proposals 99, 213-214
proposed 27, 48-49, 139, 147, 203, 251
protect 70, 115
protected 234
protection 104, 107
protocols 188
proved253
provide 18, 69, 105, 111, 134, 140, 146, 159, 182, 184, 189, 235, 244-245
provided 2, 13, 101, 141, 159, 173, 179-180, 187, 198, 213, 218, 234
provider 217
providers 86
provides 165
providing 93, 134, 185
public 135
publisher 3
pulled 114
purchase 9, 174, 214, 231
purchases 258
purchasing 1-2, 257
purpose 4, 11, 125, 151, 171, 184, 223, 228, 234-236, 243
pursued 64
pursuing 2
pushing 111
qualified 43, 64, 67, 70, 73, 249
qualifies 59, 61
qualify 52, 58, 67
qualifying 177
quality 6-7, 11, 25, 51-52, 54, 61, 65, 71-72, 78, 94, 121, 131, 135, 137, 141, 150-151, 159-160, 174, 178-180, 187-191, 195-196, 198, 212, 228-229, 232, 263
quantified 94
quantify 52
quantity 258
question 12, 17, 28, 44, 58, 74, 91, 104, 106, 192
questions 9-10, 12, 176, 200, 239
quickly 11, 61, 63
quotations 258
radically 70
raised 160
raising 137
rather 117, 262

ratings 213
rational 157
rationale 142, 227
reached 20
reaching 105
readings 93
realistic 20, 61, 110, 140, 178, 205-206
Reality 236
realize 2, 45
realized 126
realizing 1
really 9, 25, 30, 145, 194
reason 122, 124
reasonable 83, 118, 139, 142, 211
reasons 35, 142, 173
reassigned 185
rebuild 108
recasts 181
receive 10-11, 34, 47, 199, 214, 231
received 37, 114, 173, 257, 264
receiving 229
recently 107
recipient 21, 259
recognised 84
recognize 4, 17-21, 76, 79, 88
recognized 18, 20-22, 24, 73
recommend 108, 114, 150
recorded 258
recording 3, 223
records 59-60, 105, 188, 245, 260
recovery 52, 152
recruit 188
redefine 20, 34
re-design 61
reduce54-55, 137, 173, 187, 193, 228
reduced 81, 209
reducing 96, 117
referenced 190
references 265
reflect 71, 93-94, 101
reform 55, 99, 109, 121
reforms27, 49-50
refreshed 2

refuse 230
refuses 209
regarding 115-116, 145, 221
Register 4, 6, 134, 203, 207, 239, 250
regular 37, 39, 73, 159, 249, 251
regularly 41, 131, 150
regulation 207
regulatory 19
reimbursed 233
reinforce 173, 187
reject 217
rejected 223
relate 65, 225
related 17, 53, 65, 94, 150, 186, 193, 199, 239, 246-247, 262
relation 26, 79, 125
relations 113
relative 76, 86, 99, 237
relatively 119
release 141, 159, 198, 262
releases 251
relevant 35, 55, 69, 86, 97, 114, 151, 195, 217-218, 234, 258
reliable 209
relieved 2
relying 236
remain 29, 263
remaining 177, 183
remunerate 87
renewed 174
repair 141, 179, 211
repeatable 201
rephrased 11
replace 47, 132
replacing 145
Report 7-8, 74, 93, 165, 221, 229, 243, 251, 263
reported 145, 150, 156, 239, 247
reporting 68, 95, 111, 141, 145, 150, 157, 202, 215, 223, 243
reports 2, 47, 101, 134, 140, 145, 150, 156, 179, 181, 215, 246
represent 77, 225-226
reproduced 3
reputation 120
request 7, 186, 213, 223, 225-226
requested 3, 87, 223, 225
requests 197, 214, 223-224

require 32, 48, 64, 72, 98, 102, 133, 169
required 22-23, 31, 33, 38, 41, 54, 59, 84-85, 88, 130, 142, 145-146, 152, 163-164, 173, 176, 178, 191, 198, 200, 205, 217, 230, 249, 257, 262
requiring 134, 259
research 17, 108, 126, 173, 262
reserve 158, 179
reserved 3
reside 77, 181
Resistance 217-218
resolution 69, 83, 150
resolve 23, 232
resolved 160, 256, 263
resource 5-6, 127, 152, 156, 164, 169, 171-172, 174, 197, 201, 215, 222
resourced 230
resources 2, 4, 9, 17, 19-20, 26, 28-29, 50, 64, 85, 94, 99, 107, 110, 127, 130, 135, 141, 144, 160, 163-164, 171-172, 183, 185-187, 194, 198, 206, 220, 229, 232, 234, 242, 246
respect 3, 49
respond 137, 203
responded 13
response 17-18, 93, 97-98, 100, 102, 201, 252
responses 77, 203
responsive 176, 183
result 71, 77, 80, 151, 182, 184-185, 199, 218, 225, 244, 259
resulted 97
resulting 69, 140, 197
results 10, 31, 42, 69, 74, 76-77, 79, 82, 85, 87, 89, 96, 101, 138, 159, 183, 191, 219, 250, 253
retain 104, 188
retention 53
retrospect 114
return 80, 120, 167, 192
revenue 53
revenues 47
review 11, 132, 140, 180, 258
reviewed 42, 190, 229
reviewer 235
reviews 152, 159, 167, 237
revised 59, 97, 157, 210
revisions 259
reward 45, 57-58, 173, 221

rewarded 21
rewards 101
rework 46, 57, 187
rights 3
rigorously 229
risk-based 250
routine 97
rushing 239
safely 209
safety 76, 86, 116, 139, 249-250
samples 187
sampling 188
satisfied 118, 253, 262
satisfies 248
satisfying 123
savings 39, 45, 54, 59
scalable 78
scenario 29, 43
scenes 2
schedule 5-6, 41, 56, 91, 125, 137, 146, 159-160, 167, 173, 177-180, 193-194, 198, 201, 211, 219, 223, 225, 231, 245, 251, 261
scheduled 141, 144, 179, 211, 245, 257
schedules 167, 246
scheduling 160, 198, 215, 243, 245
scheme 102
science 65
Scorecard 4, 13-15, 220
scorecards 101
Scores 15
scoring 11
screening 139
scripts 152
seamless 117
second 13
secret 2
secrets 1
section 13, 27, 43, 57, 73, 89, 102-103, 127-128
sections 153
sector 247
secure 142
securing 54, 109
security 26, 85, 95, 98, 134, 189, 225, 230, 249
segmented 30

segments 29, 112
select 64, 99
selected 80, 183
selecting 67, 107, 139
Selection 7, 213
selects 237
self-help 2
sellers 3
selling 126, 166, 227
senior 93, 111, 115
sensitive 39, 53, 137
sequence 162
sequenced 160
sequencing 109, 180
series 12
Service 1-4, 9, 52, 86, 108, 151, 166, 186, 201, 214, 221-222, 264
services 3, 32, 50, 53, 57, 105, 113, 135, 234, 242, 251-253, 259
session 161
setbacks 61, 63
setting 110, 112
Several 72, 208
severely 61
shared 97, 137, 147, 183
sharing 81, 98
shifts 18
shopping 1
should 9, 19, 23, 28-29, 33, 45, 47, 50, 59, 64-65, 69, 71, 75, 80, 84, 94-95, 114, 116-119, 124-126, 130-131, 133-135, 137, 144, 150, 161, 163, 165, 171-172, 182, 195-196, 203-204, 206-207, 213-214, 224-225, 231, 241, 250
-should 186
signature 110
signatures 169
signers 260
similar 31, 40, 69, 71, 82, 162, 173, 185, 194, 211
simple 119, 236, 247
simply 10, 228
single-use 9
situation 2, 18, 44, 182, 204, 206, 247
situations 92
skeptical 116

skills 22, 26, 109-110, 165, 181, 185, 197, 217, 219, 235, 238, 243-244
smallest 22, 80
soccer 1
social 116, 135, 209
societal 125
software 22, 65, 141, 144, 158, 173, 178, 201-202, 205-206, 211, 217, 251
solicit 35, 233
solution 1, 49, 69, 74, 77-80, 83, 86-87, 89, 91, 151-152, 241, 251
solutions 48, 74-75, 79, 86, 95, 263
solved 23
solving 235
Someone 9
someones 235
something 118, 185, 187, 194
Sometimes 48
source 7, 116, 121, 181, 209, 213
sources 29, 61, 71, 233
special 38, 101, 132
specific 10, 23, 35, 40, 42, 66, 106, 162, 164, 166, 168-169, 174, 178, 181-182, 195, 211, 218, 222, 225, 243, 255
specified 106, 157, 159, 255, 258, 261, 264
Speech 146
spending 2
spoken 107
sponsor 24, 142, 151
sponsors 21, 192, 231
spread 92, 97
stable 207
staffed 29
staffing 22, 101, 152
standard 9, 169, 197, 251
standards 11-12, 92, 94-96, 101, 152-153, 187-188, 190, 198, 215, 225
started 10, 165
starting 11
starts 138
stated 107, 115, 193-194, 196, 255
statement 5, 12, 84, 86, 150-151, 157, 174, 182, 186
statements 13, 27, 35, 41, 43, 57, 71, 73, 89, 103, 128, 174, 192

status 7-8, 65, 141-142, 150, 179, 194, 198, 205, 211, 221, 223, 239, 246-247, 251
steady 50
Steering 152, 216
stopper 147
storage 224
stored 230
stories 30
straight 258
strategic 47, 84, 99, 121, 216
strategies 75, 92, 117, 121, 153, 197, 208, 228
strategy 23, 42, 53, 75, 83, 87, 96, 107, 114, 116-117, 138, 228, 249, 258, 262
Stream 65, 67
strength 230
strengths 159, 165, 233, 253
stretch 112
strict 64
strive 112
Strongly 12, 17, 28, 44, 58, 74, 91, 104
structure 5, 54, 81, 114, 119, 154, 171, 173, 178, 211, 217
structured 127, 141
structures 245
stubborn 127
stupid 125
subdivided 194
subject 10-11, 31
subjects 70
submission 257
submitted 225
subset 22
sub-teams 235
succeed 112, 236
success 23, 26, 29, 32, 39-40, 45-46, 55, 57, 79, 84-85, 96, 108, 112-113, 120, 123-124, 126, 130, 132, 168, 173, 187, 189, 198, 219, 221, 227, 236, 256, 263
successes 112
successful 1, 60, 75, 77, 101, 108, 123, 131, 221
succession 95
sufficient 137-138, 223, 230, 238
suggested 98, 225-226
suitable 202, 205, 231
summary 156, 246

Sunday 1
superior 1, 197
supervisor 234
supplier 84, 126, 159, 205, 229, 258
suppliers 33, 65-66, 180, 257-258
supplies 253
supply 50, 257
support 3, 9, 20, 63, 87, 91, 95, 101, 105, 115, 117, 133, 146, 169, 196, 201, 218, 228, 230, 246, 252
supported 72, 205
supporting 84, 93, 179
supportive 195-196, 211
supposed 229
surface 98
SUSTAIN 4, 87, 104
sustaining 102, 166, 234
symptom 17, 57
system 11, 36, 67, 99, 107, 119, 132, 144-146, 151, 158, 171, 187, 202, 225, 228-230, 245, 248
systematic 51
systems 1, 58, 62, 65, 71, 82, 85, 100, 141-142, 173, 180, 219, 247, 254
tackle 57
tactics 228
tailored 2
taking 53, 131, 221, 257
talent 73, 111
talents 109
talking 9
tangle 194
target 29, 141, 218
targets 112, 131, 247
tasked 99
technical 88, 140, 165, 213, 238, 241
techniques 70, 116
technology 1, 47, 78, 108, 166, 233, 250
templates 9-10
tender 258
tenders 257
test-cycle 191
tested 25
testing 158, 189, 205, 215, 250, 263
themselves 1, 235

therefore 210
things 76, 122, 131, 150, 182, 201, 206, 220, 255-256
thinking 68, 79, 122
third- 86
thorough 225
thought 200
threat 21, 105
threats 1-2, 210
through 61, 73, 150, 227
throughout 3, 63, 122, 153, 167
Thursday 1
tighter 115
time-bound 35
timeframe 62, 183
timeframes 23
timeline 189, 225, 241
timely 22, 80, 160, 220, 223, 245
Timescales 136
timetable 167
Timing 205, 231
together 108
tolerable 210
tolerances 75
tolerated 175
tomorrow 96
toolkit 1-2
toolkits 1-2
top-down 97
topics 88
touched 146
toward 96, 221
towards 2, 69
traced 152, 185
tracked 150, 180, 197
tracking 41, 99, 152, 178, 211
traction 116
trademark 3
trademarks 3
trade-offs 216
trained 36, 206
training 19, 22, 25, 27, 80, 92, 101, 217-218, 221, 230, 234, 237-238, 264
trainings 21

Transfer 13, 27, 43, 57, 73, 90, 99, 101, 103, 128, 251
translated 42
travel 111, 234
trends 63, 66, 69, 89, 125, 139-140, 188, 209
trigger 75
triggered 160
triggers 76, 160
trouble 104
trying 9, 112, 127, 209, 218, 229
typical 130
ubiquitous 127
unaware 1
unclear 33
uncovered 2
underlying 88
undermine 127
understand 36, 71, 131, 219, 236, 238, 257
understood 82-83, 114, 205, 208, 244
undertake 60
underway 87
uninformed 117
unique 2, 111, 166, 186
Unless 9
unplanned 234
unprepared 1
unpriced 193, 246
unproven 201, 206
unresolved 169
update 1, 185
updated 10-11, 71, 167, 174, 179
updates 11, 101, 251
upfront 217
up-sell 113
up-to-date 250
usability 88
useful 85, 94, 153, 201, 264
usefully 11, 22
UserID 165
usually 1
utility 175
utilizing 1, 78
vacations 257
validate 56, 248

validated 33, 35, 42, 63, 73
Validation 248
valuable 9
values 93, 209, 258
variables 58, 97, 173, 227, 237
variance 8, 235-236, 245
variances 142, 156-157, 193, 211, 245
variant 257
variation 17, 31, 61, 71, 96, 160
vehicle 76, 86
vendor 76, 201, 206
vendors 26, 67, 86, 180, 215
verified 11, 33, 35, 42, 63, 142, 166
verify 45, 47-56, 94-95, 99, 245, 248, 259
verifying 56
Version 251, 265
versions 29, 37
versus 144
vested 119
viable 95, 154
vigorously 235
virtual 237
vis-à-vis 206
vision 139-140, 196
visualize 161, 175
voices 134
volatile 88
volatility 202
waited 2
walking 2
warning 160
warrant 232
warranty 3
weaknesses 142, 165, 214, 253
website 224
weekly 131
weeknights 1
wellness 173
whether 9, 92, 124
wholesaler 159
widespread 93
willful 250
willing 189, 249-250

within 1-2, 60, 87, 161, 163, 209, 221, 226, 237, 253
without1, 3, 13, 108, 124, 206, 225, 236, 259
worked 140, 194, 242
workers 113
workflow 227-228
workforce 22, 89, 112, 123
working 2, 92, 101, 152
Worksheet 6, 175, 183
worst-case 29
writing 148
written 3, 179, 229
yesterday 21
youhave 161
yourself 110, 112, 151, 195

Made in United States
North Haven, CT
03 November 2026

21727484R00170